AF855746

BIBLE BACKGROUND

GEOGRAPHY, HISTORY,
AND DAILY LIFE OF BIBLE TIMES

Glen L. Thompson

Glen L Thompson

Bible Background – Geography, History, and Daily Life of Bible Times
Glen L. Thompson

Published in cooperation with:
Asia Lutheran Seminary
8/F., Chun Hoi Commercial Building, 688 Shanghai Street,
Mong Kok, Kowloon, Hong Kong.

For further information:
Email: info@als.org.hk
Website: www.als.orghk

Asia Lutheran Seminary Theological Studies – Volume 3
Chinese Edition (2015) ISBN: 978-988-14167-2-8
1st English Kindle Edition (2016) ISBN: 978-988-14167-5-9
English Kindle paperback Edition (2019) ISBN: 978-988-14167-3-5
English Hardcover Edition (2022) ISBN: 978-988-14168-8-9

All rights reserved ©2022 by Glen L. Thompson

Table of Contents

LIST OF MAPS

LIST OF CHARTS

GETTING STARTED

This textbook is designed to guide Bible students through a brief overview of Bible history while at the same time teaching them the basics of geography and daily life in Bible times. It is hoped that this kind of study will help a Bible student fill in the gaps in his knowledge. It provides him with the "big picture" and flow of biblical events. It will also acquaint him with the historical situation and day-to-day life in ancient times so that the Bible will come alive as it is read or taught. The text was originally written for readers whose native language was not English, so the prose and vocabulary have been kept as simple as possible.

In the first five lessons the student will learn how to construct a simple map showing the main geographical features of Palestine. Lessons 6-37 will lead the student chronologically through Bible history using the Bible map to locate important places and events. Interspersed with the Bible history lessons are chapters on other ancient nations, studies on food, clothing, occupations, trade and money, and chapters on special topics of relevance, such as the Hebrew calendar, weather, the geography of Jerusalem, the origin of Pharisees and Sadducees, and the Apocrypha. A series of worksheets is available to accompany the text. The worksheets will help the student interact with the text and review its main points.

Except for liters (which are now commonly used worldwide), measurements are given using traditional English measures (feet, yards, miles), with metric conversions added in parentheses. Bible references in my original English text were to the 1984 NIV. When key terms are no longer used in subsequent revisions of the NIV, I have tried to note that fact and refer to other translations which do preserve that term.

PDF versions of the maps and charts can be downloaded free of charge at www.als.org.hk/bible-background.

ACKNOWLEDGMENTS

This text was first written during the 1980-81 school year for use at the Lutheran Seminary in Lusaka, Zambia. The author had just returned from six weeks of traveling in Israel and taking part in the excavations at Tel Michal. From its inception, the course was built around a Bible map which students are taught to draw—one of many great pedagogical tools developed by Dr. John C. Lawrenz. I thank him for allowing me to use it as the basis of this course.

In 1998 a second edition of the course was produced for use in the WELS Multi-ethnic Pre-Seminary Program. The present version is a revision of the course used in recent years at Asia Lutheran Seminary in Hong Kong.

The translation and publication of a Chinese edition in 2015 made the author aware that a more thorough updating was needed. That was done in 2016 with the help of Nathanael Jensen (text), Luke Thompson, and Ivy Choi (graphics). I sincerely thank them for their efforts.

Glen L. Thompson
Hong Kong
October, 2016

LESSON 1:

THE WORLD OF THE BIBLE AND MAKING A BIBLE MAP

(Steps 1-5)

In this book we will be studying three related subjects: 1) basic Bible history; 2) the geography of the Holy Land and the Middle East; and 3) daily life and customs during biblical times. In order to better understand the Bible we want to first gain a good grasp on the geography of Bible lands. In this first lesson we want to learn about other areas surrounding the land of Israel, and start to make our own map of the Holy Land using a regular A4 (or 8.5 x 11) sheet of paper, a pencil and a ruler.

At the end of this lesson you should be able to:

* *Briefly explain in what ways Israel lay in an important place in ancient times;*
* *Give a short definition of the term "Fertile Crescent;"*
* *Locate the Fertile Crescent on a map; and*
* *Properly complete Steps 1-5 of making a Bible map.*

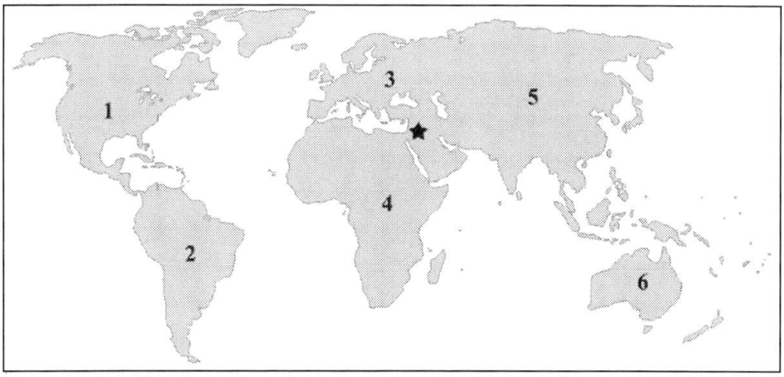

Map 1. Israel's Location in the World

A. THE LAND OF ISRAEL AND THE WORLD AROUND IT

Map 1 shows our world with the continents numbered: 1. North America, 2. South America, 3. Europe, 4. Africa, 5. Asia, and 6. Australia.

The large star indicates the land of Israel. Notice that Israel is strategically located between the continents of Europe, Africa and Asia.

B. ISRAEL AND THE ANCIENT NEAR EAST

In accordance with God's plan and purpose, the nation of Israel was situated in a very important place. The area in which ancient Israel was located is commonly known today as the Near East or Middle East. Map 2 gives us a close-up picture of the Eastern Mediterranean and the Near East. It reminds us that this area of the world was where many of the key events of early history took place. It is probable that the Garden of Eden and other places mentioned in the early chapters of Genesis were once a part of this area. It is in this area of the world that most of the events of the Bible also took place. Note that Israel is also at the center of the Near East.

On Map 2 you can see the names of the most important areas of Bible times. Mesopotamia (often referred to as the cradle of civilization) was home to the powerful kingdoms of Sumeria, Babylon, and Assyria. Egypt was already a great nation 2,000 years before Christ was born. The area known as Syria (much larger and more important than modern day Syria) would later be controlled by the western kingdoms of Greece and Rome. These two latter groups had a significant influence on the events of the Near East already several centuries before the coming of our Savior and this lasted throughout the centuries following Christ's resurrection as the Christian church was being established.

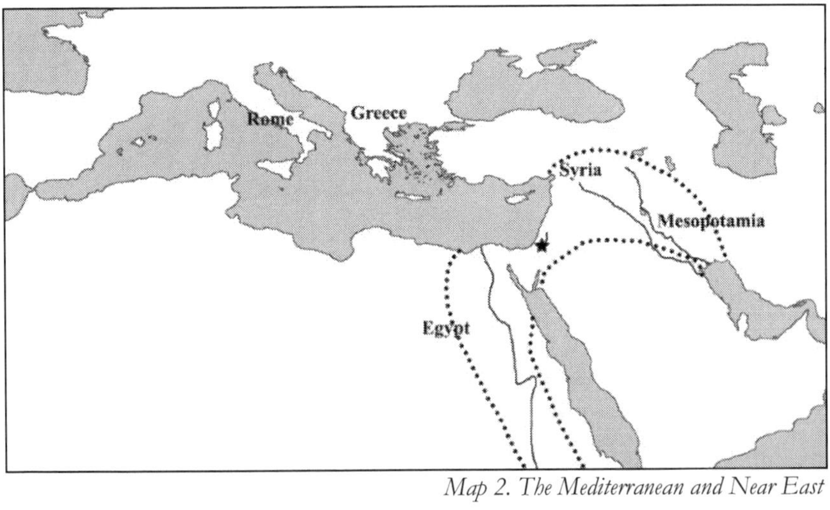

Map 2. The Mediterranean and Near East

In ancient times these powerful nations interacted with one another. Often this interaction was in the form of trade, but on many occasions, it was the result of warfare. But whether these nations were trading or fighting with each other, they usually had to travel through Israel. With traders and armies both traveling the highways of Israel this small country certainly had a place of importance in the ancient world. Perhaps this was one reason why God led Abraham to the land of Israel and gave this land to his descendants. Thus, whenever the traders and armies marched through Israel, they could have an opportunity to hear about the God of Israel. God wanted all these nations to know him and his will for them just as he had revealed it to Israel. God wanted his people, the children of Israel, to lead lives that reflected the gospel message and to verbally share his truths. Israel was also to be an example to other nations of proper worship of the true God. The children of Israel had a wonderful opportunity to accomplish this because of the strategic location in which God placed them and their nation.

There is one additional feature that we should note about the area of the Near East. Note the dotted area of the map extending from Mesopotamia through Syria and into Israel. This crescent or moon-shaped area is often referred to as the Fertile Crescent. It was very fertile land, good for growing a variety of crops. Outside of the Fertile Crescent the land of the Near East mainly consists in mountains, sparse vegetation, and desert. Thus, the area of the Fertile Crescent became a very valuable piece of property. Powerful countries tried to conquer this area so they could better support their populations. Because there was much more water and food in the Fertile Crescent, many large cities were located in the region. Again, note that Israel was carefully placed within the area of the Fertile Crescent not only to enjoy its rich resources but also to play a key role in its development. God had carefully selected a special place for his people and their welfare.

C. THE MAKING OF A BIBLE MAP – STEPS 1-5

To properly understand the Bible, it is important to understand the geography of Israel. To assist you in doing this you will be learning how to draw your own map of the Holy Land. In the first five chapters you will learn to draw a map of Palestine. This will help you locate Israel's most important lakes, rivers, valleys, mountains, etc. When you are done you

should be able to take a blank sheet of paper (either 8.5 x 11 inch or A4 sizes will work) and pencil and draw this map by yourself.

1. First, take your sheet of paper and fold it in half three times. Then unfold your paper and you should have eight equal vertical segments.

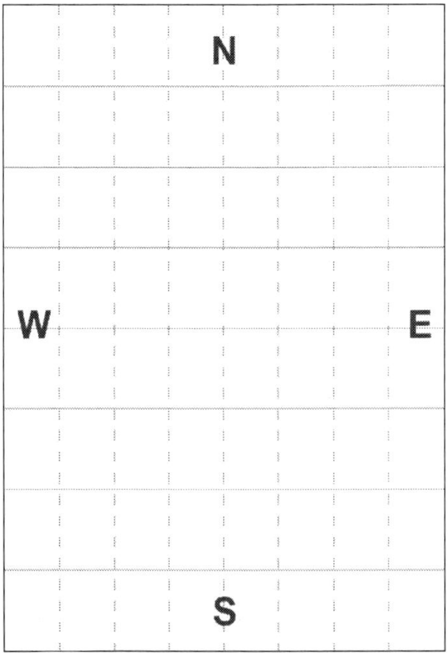

2. Then, take your sheet of paper and fold it in half three times in the other direction. Once you have finished this and have unfolded your paper, you should have eight equal horizontal segments which bisect the previous folds forming an 8x8 grid. At the middle crease on the top of the page place an N for north; at the middle fold at the left of the page place a W for west; opposite of it, on the right of the page place an E for east; and at the middle fold on the bottom of the page place an S for south.

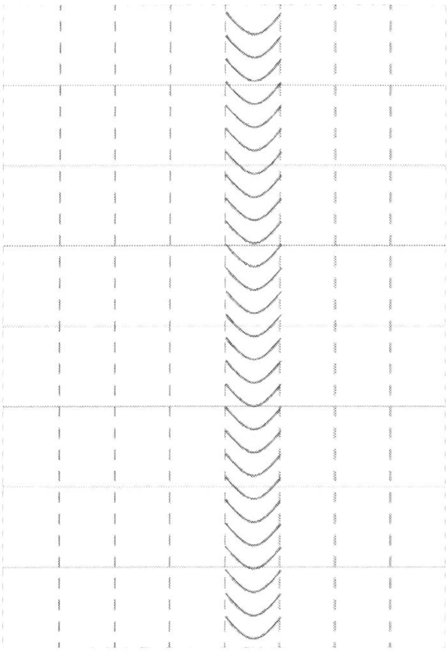

3. In column 5 lightly draw a series of bent lines extending from the top of the page to the bottom to indicate the Rift Valley. This deep valley was often called the Arabah in the Old Testament (Deuteronomy 3:17; Joshua 12:3). It is part of the larger Rift Valley that extends through the Red Sea all the way into central Africa. It is so deep that it can be seen from space!

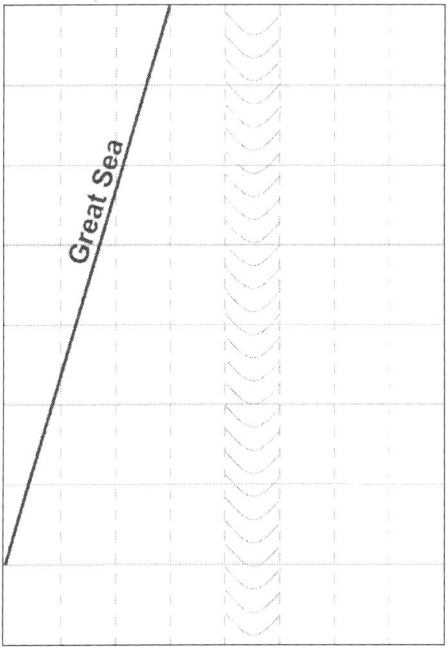

4. Draw a line from the third fold at the top of the page to the last fold at the left side of the page (as shown in the diagram). This line indicates the seacoast. Label the area to the left of this line the Great Sea; today we call it the Mediterranean Sea. When God described to Moses the land he was giving to his people Israel, he said "Your western boundary will be the coast of the Great Sea" (Numbers 34:6). The Great Sea is mentioned numerous times in the Old Testament. It could also be referred to as the Western Sea (Deuteronomy 34:1; Joel 2:20), or the Sea of the Philistines (Exodus 23:31). Although the Hebrew Old

Testament uses these different terms, modern translations usually translate all these passages as the Mediterranean Sea.

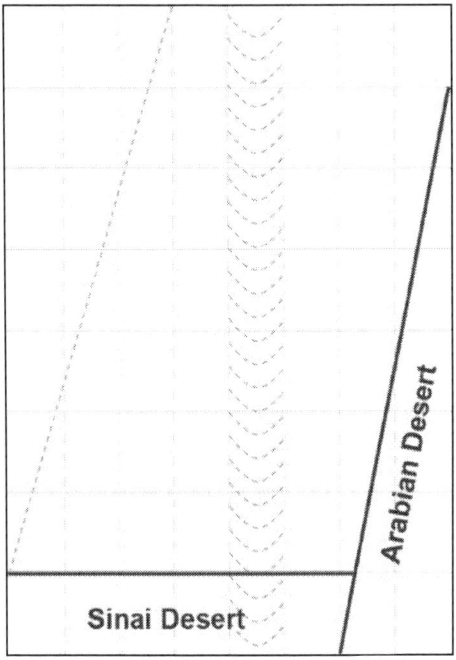

5. Draw a line from the first fold on the right side of the page to the second fold from the right on the bottom edge of the page (as shown in the diagram) to indicate the edge of the desert. Label the area to the right of the line as Arabia, or the Arabian Desert. The Israelites traveled along this line on the edge of the desert as they went from Egypt back to the land of Israel (Judges 11:18). Caravans of traders would carry their goods along the edge of this Arabian Desert (Isaiah 21:13). Some nomadic tribes lived in the desert, ruled by their sheiks or kings (Jeremiah 25:24). Along the bottom fold of the paper draw a line from the seacoast to the edge of the desert. Label the area below the line as the Sinai Desert. This area extends far off the bottom of our map. The Israelites spent much of their forty years of wandering in this desert (Exodus 19:1; Numbers 1:1, 33:16). It is where Mount Sinai is located. This desert area could also be considered part of the Arabian Desert (Galatians 4:25). The Lord told Joshua, "Your territory will extend from the desert to Lebanon, and from the great river, the Euphrates-- all the Hittite country-- to the Great Sea on the west (Joshua 1:4). The desert he is referring to here includes both the Sinai and the Arabian Desert which together formed the southern and eastern boundaries of the Promised Land.

We have now finished steps 1-5 of drawing our map. Your map should look like Map 3. Practice steps 1-5 until you can draw and identify the various geographical features without the instructions.

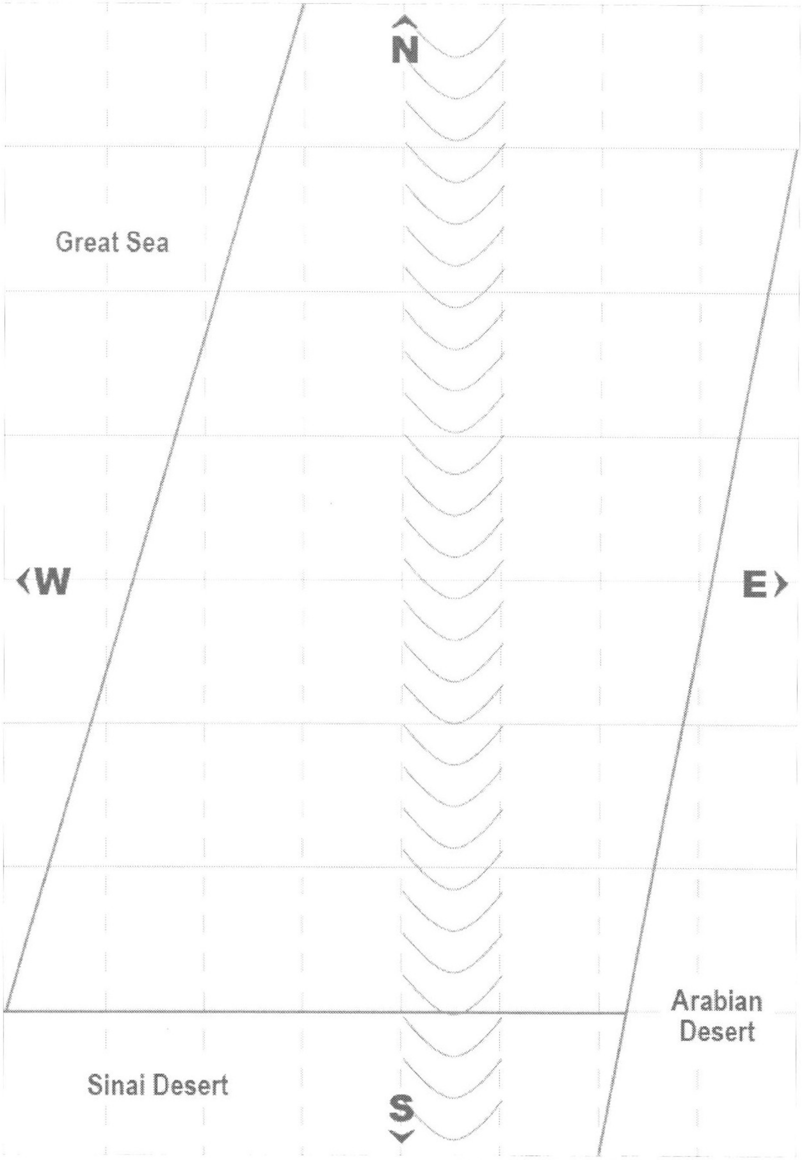

Map 3. The Bible Map, Steps 1-5

LESSON 2:
WATER AND LIFE IN THE NEAR EAST AND MAKING A BIBLE MAP
(Steps 6-13)

In Lesson 1, we learned that Israel was placed in an important area called the Near East. In this lesson, we want to learn more about the Near East. Especially, we want to learn about the close connection between water and life in the Near East. We will also be learning about the main sources of water in Israel, and we will be putting them on our Bible map.

At the end of this lesson you should be able to:

* *Name and locate on a map four major rivers of the Near East;*

* *Name and place on your Bible map the seven main watercourses of Israel; and*

* *Briefly explain the Near Eastern proverb "Water is life."*

A. WATER IN THE NEAR EAST

In Lesson 1, we learned about the Fertile Crescent. This area of land in the Near East was called the Fertile Crescent because it was well supplied with water and was shaped like a crescent moon. Four important rivers supplied much of the water for the Fertile Crescent.

Two of these major rivers are in Mesopotamia. Look again at Map 2 and you will see the lines marking these rivers. The name Mesopotamia means "between the rivers." The region was given this name because many of the major cities were located between these two important rivers. The first river of note is called the Tigris River (Genesis 2:14; Daniel 10:4). This is the river closer to the top and right of the map. The important cities of Nineveh, Ashur, and Baghdad were on this river. The second river is farther towards the bottom and left of the map. It is called the Euphrates River and was also referred to as "The Great River" (Genesis 15:18; Revelation 9:14, 16:12). The famous cities of Babylon and Ur were located along this river.

The third river is in Africa. It starts far inland in the mountains of Africa and after flowing through Egypt enters the Mediterranean Sea. It is called the Nile River. Almost all the people of Egypt lived along the banks of this river. Every year after the rainy season in the mountains of Africa, the Nile floods its valley (Jeremiah 46:7; Amos 8:8). When the flood water goes down, it leaves behind rich soil for farming. The fourth river is much smaller than the other three. It is not important because it is big or has important cities built along it. It is important because it flows through the middle of Israel. This is the Jordan River.

Around these four rivers, the earliest civilizations of the Near East grew up. Why did nations arise around rivers? One of the most important things that people need to live is water. Water is always one of life's basic necessities. People need water to drink, wash, and grow their crops. People need water also for their animals. Wild animals could be hunted near the rivers where they came to drink. Likewise, fish were also an important food source. For all of the above reasons, people usually settled where there was plenty of water.

In addition, the weather for most of the year is very warm in the Near East. Rain comes only during the rainy season. Some narrow valleys turned into streams after rainstorms, but these would soon dry up again. These were called *wadis* (Deuteronomy. 21:4 [NET]; 1 Samuel 15:5 [NET]). In other places there were natural watering holes. During the dry season, there was often a shortage of water. Wells dried up. Water storage tanks became empty, and all water became very precious. The large rivers of the Fertile Crescent did not dry up, however. They became smaller, but some water always remained in them. Farmers also dug small canals to take the water from the river into their fields. This is why the earliest settlements tended to be near these large rivers.

In the Near East there is a proverb: "Water is life." In other words, when you have no water, you die. Only when water is found in sufficient abundance can life exist. The four major rivers of the Near East provided water throughout the year; for this reason, they became the center for life in this area.

B. MAKING A BIBLE MAP – STEPS 6-13

In Lesson 1, we learned the first steps in making a Bible map. Take the map that you made and let's continue to work on it. In steps 6-13, we will add five rivers and two seas to our map. They are the seven main water sources in the land of Israel.

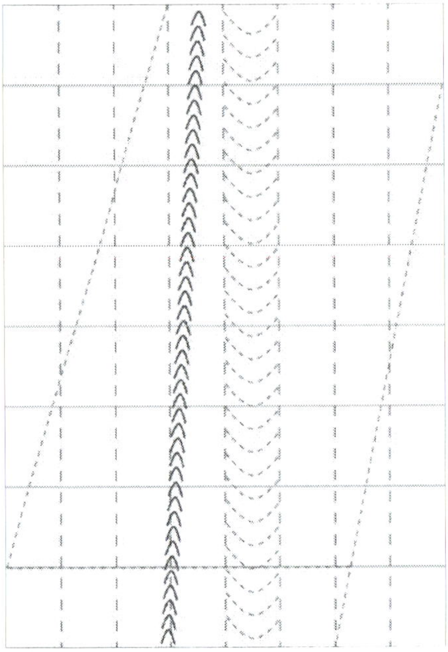

6. At the top edge of the paper, find the middle point between the edge of the sea and the Rift Valley. Draw a series of small mountains from there to the third fold from the left edge of the paper on the bottom. This line signifies the Watershed Line. The watershed was the ridge of high ground that determined which way the rivers flowed. All rivers to the west of this line flowed into the Mediterranean Sea. All rivers and streams to the east flowed into the Rift Valley and eventually the waters end in the Dead Sea.

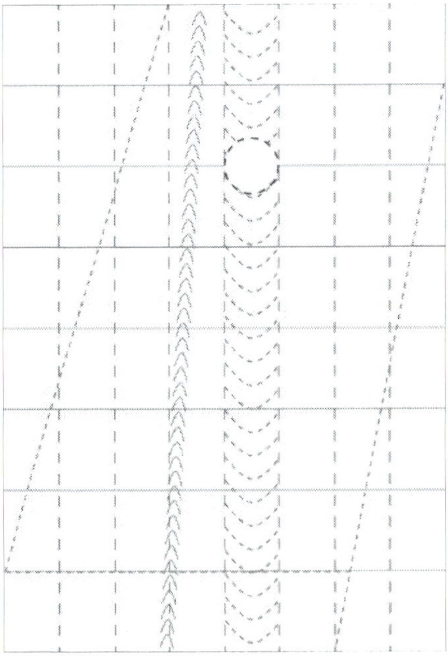

In the previous lesson we have already identified the strip that goes down the middle of our map as the Rift Valley or Arabah (Joshua 12:3). The Arabah contained the three most important bodies of water within the land of Israel.

7. Draw a circle from one edge of the Rift Valley to the other on the second horizontal fold from the top. This circle represents the Sea of Galilee. Jesus often walked and preached along the shores of this lake (Matthew 4:18; 15:27). In Old Testament times it was sometimes called the Sea of Kinnereth (Deuteronomy. 3:17; Joshua 11:2). The word "kinnereth" may come from the Hebrew word for a stringed musical

instrument which had the same shape as the lake. (This became Gennesaret in the New Testament, as in Luke 5:1.) In New Testament times it was also at times referred to as the Sea of Tiberias because of the new city by that name along its shore (John 6:1; 21:1). This lake is about 8 miles (13 kilometers) wide and 13 miles (20 kilometers) long. It lies about 700 feet (215 meters) below the level of the Mediterranean and is up to 140 feet (43 meters) deep. It is the only large body of fresh water in the land of Israel. Storms can appear without warning, such as the one that Jesus calmed when he was on the sea with his disciples (Matthew 8:23-27).

8. Draw an oval the width of the Rift Valley between the first and the third folds from the bottom. This oval signifies the Dead Sea. In Hebrew Bible times this was usually called the Salt Sea (Genesis 14:3, Numbers 34:3; Joshua 15:2). It contains a very high concentration of minerals, and so it tastes very salty, and fish cannot live in it (which is why it is today called the Dead Sea). It was sometimes also called the Sea of the Arabah, since it was at the bottom end of the Arabah or Rift Valley (Deuteronomy 3:17; Joshua 12:13).

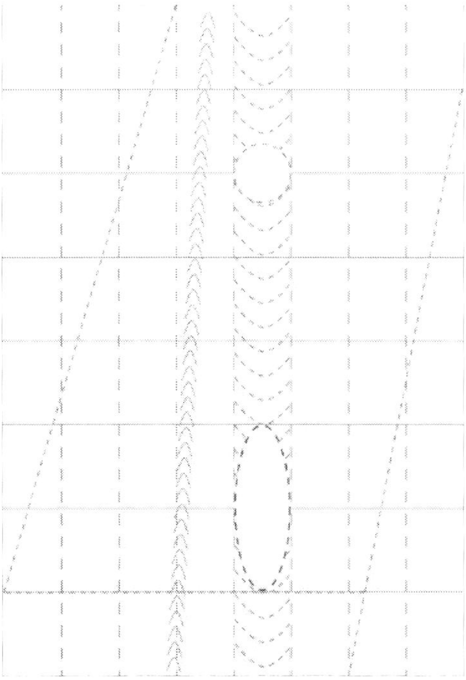

The Dead Sea was much larger than the Sea of Galilee, about 9 miles (15 kilometers) wide and 31 miles (50 kilometers) long. The Rift Valley is deepest at the Dead Sea, with the shore being 1,300 feet (400 meters) below sea level, and the Dead Sea is another 1,000 feet (300 meters) in depth.

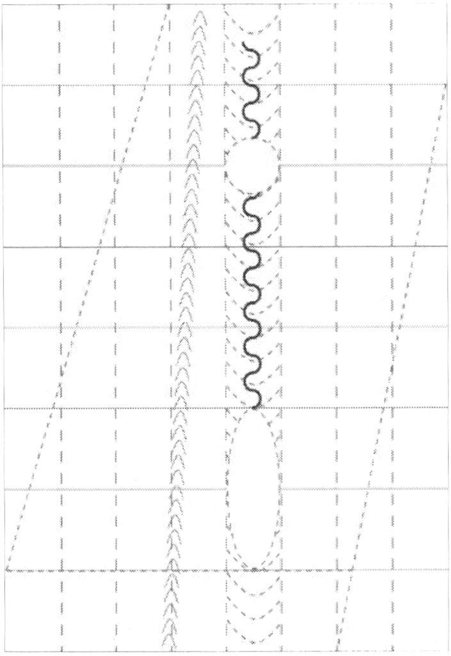

9. Draw a wavy line in the Rift Valley from the Sea of Galilee to the Dead Sea. Draw a similar line above the Sea of Galilee as shown on the diagram. This line signifies the Jordan River. The Jordan River is about 150 miles (240 kilometers) long—about 60 miles (100 kilometers) above the Sea of Galilee and about 75 miles (120 kilometers) long between the Sea of Galilee and the Dead Sea. The river begins with several sources and springs flowing out of the mountains of Lebanon to the north of Israel.

During the rainy season the Jordan River became quite wide and deep (Joshua 3:15), but the Lord still stopped the water flowing so the Israelites could cross over and re-enter the land of Israel after their journey from Egypt (vv. 16-17). It was also in the Jordan River that John baptized Jesus (Matthew 3:13).

The last four bodies of water we will put on our map are all rivers flowing from the edge of the Arabian Desert down to the Rift Valley.

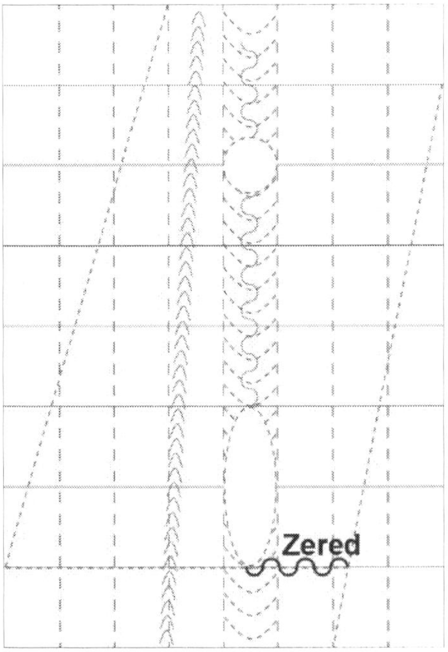

10. Draw a wavy line from the bottom of the Dead Sea (along the first fold from the bottom) to the edge of the Arabian Desert. This line signifies the

Zered River. The Israelites crossed the Zered River on their way to the Promised Land (Numbers 21:12; Deuteronomy 2:13-14).

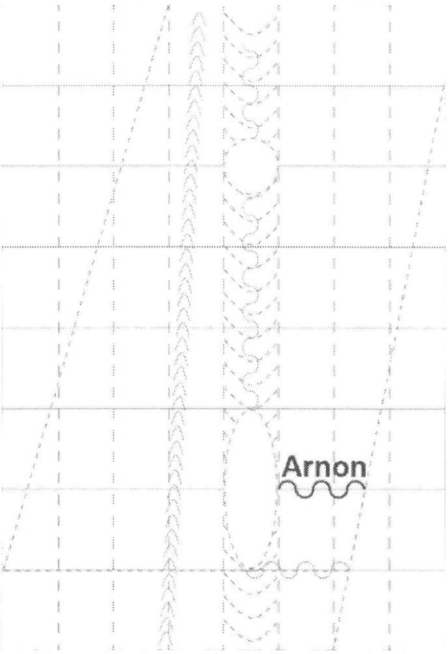

11. Draw another wavy line from the eastern edge of the middle of the Dead Sea (along the second fold from the bottom) to the edge of the Arabian desert. This line signifies the Arnon River. The Israelites also crossed the Arnon on their way to the Promised Land. It was in the land of the Moabites and Amorites (Numbers 21:13). Along most of its length, the Arnon flowed at the bottom of a deep valley, so the Old Testament often refers to the Arnon Gorge (Deuteronomy 2:24, 36; Joshua 12:1-2; 2 Kings 10:33), The few places where it was easy to cross (fords) were used by many people (Isaiah 16:2).

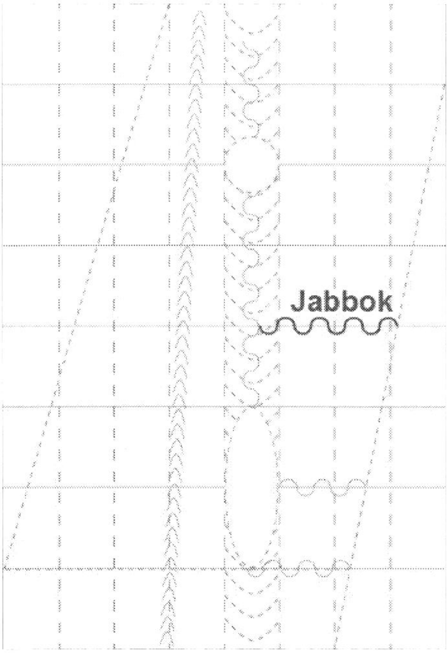

12. Draw a third wavy line on the middle fold from the Jordan River to the edge of the Arabian desert. This line signifies the Jabbok River. Jacob crossed the Jabbok River on his return from living in Syria (Genesis 32:22). The area between the Arnon and the Jabbok was later given to the tribes of Reuben and Gad (Deuteronomy 3:16).

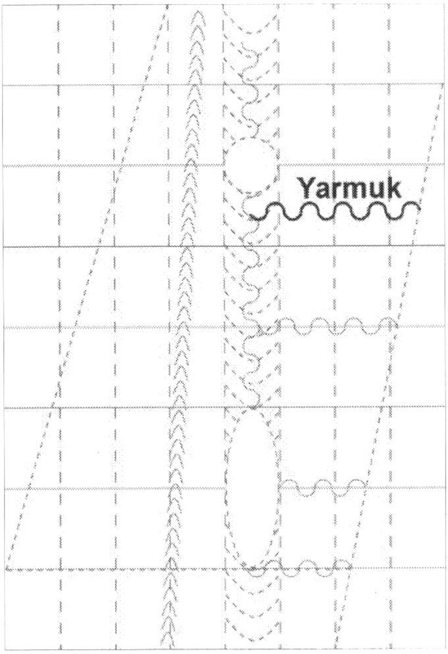

13. Draw a fourth wavy line from the Jordan River just beneath the Sea of Galilee straight east to the edge of the Arabian desert. This line signifies the Yarmuk River. The Yarmuk is not mentioned in the Bible. In Old Testament times the area around the Yarmuk was part of the land of Bashan. In New Testament times the Yarmuk was included in the areas of Trachonitis (Luke 3:1) and the Decapolis (Matthew 4:25; Mark 5:20, 7:31).

You have now finished steps 1-13 of making a Bible map. Your map should look like Map 4. You have placed the seven most important watercourses on the map. Notice that the first three are all in the Rift Valley, and the other four are all on the eastern side of the Jordan River and flow into the Rift Valley. Thus, there are no major bodies of water or rivers on the western side of the Jordan. This meant that the Israelites living there had to rely on springs and wells for drinking water, and rain and dew for growing their crops. We will learn about the Israelites' sources of water in more detail later. Practice the first 13 steps of the Bible map until you can do them without looking at the instructions or illustrations.

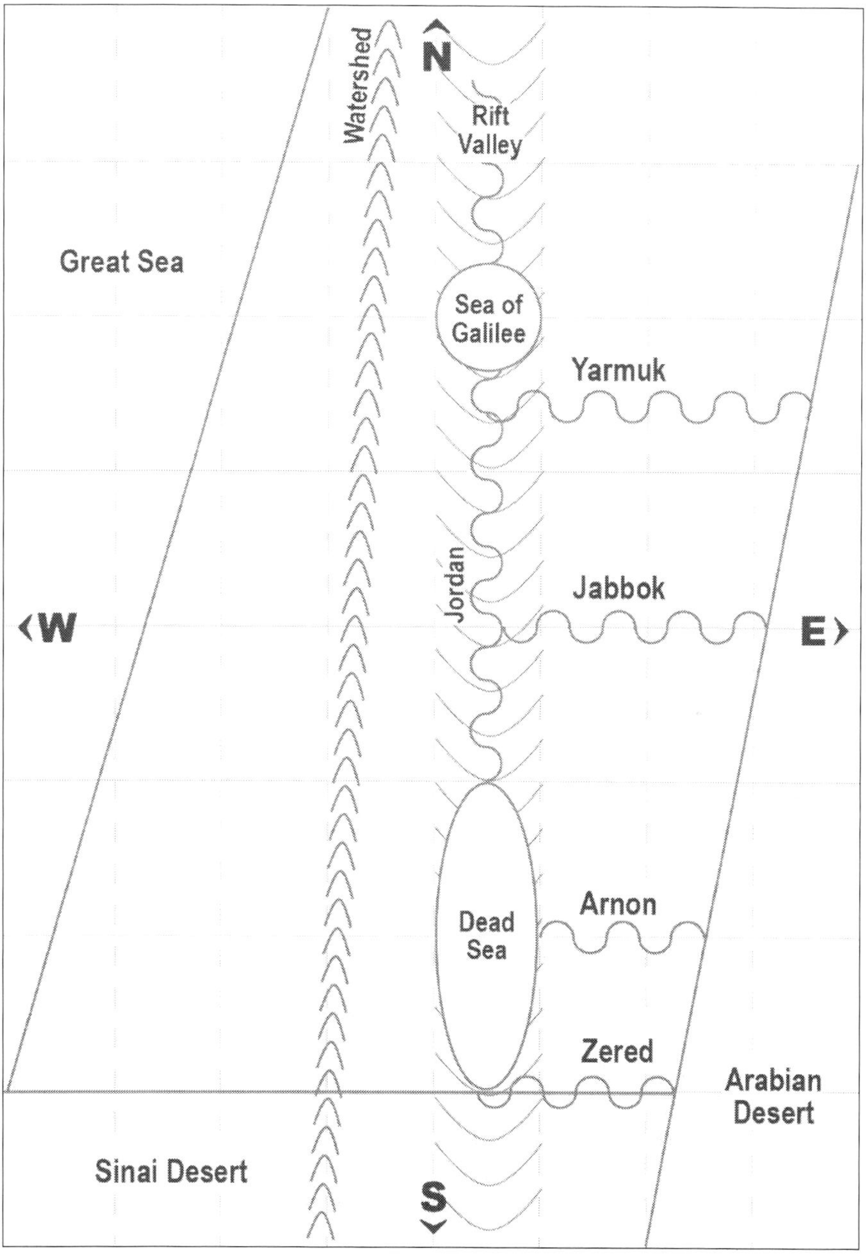

Map 4. The Bible Map – Steps 1-13

LESSON 3:
MAIN PERIODS OF BIBLE HISTORY AND MAKING A BIBLE MAP
(Steps 14-16)

In our study of the land and people of the Bible, it will be helpful if we can picture in our minds the major time periods of Bible history and their approximate dates. That is one of the goals of this lesson.

At the end of this lesson you should be able to:

* *Write the names and approximate dates of the nine main time periods of Bible history;*

* *Make a simple time line showing the nine main periods; and*

* *Properly complete steps 14-16 of making a Bible map.*

A. THE NINE PERIODS OF BIBLE HISTORY

Below is a simple time line showing the nine main periods of Bible history:

1) Creation until Abraham; 2) The Patriarchs; 3) The Captivity in Egypt; 4) The Exodus and Judges; 5) The United Kingdom; 6) The Divided Kingdom; 7) The Babylonian Exile and Return; 8) The Intertestamental Period; and 9) The New Testament Period. If you can remember the names and dates for these periods, it will help you place the smaller events of Bible history in their proper order and context. Later in this book we will have one or more lessons on each of these periods.

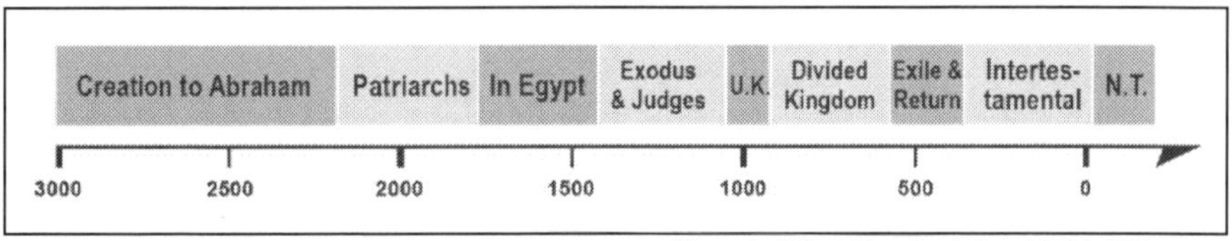

Chart 1. The Nine Periods of Bible History

1) Creation until Abraham (Creation to 2200 B.C.). This period is described in Genesis 1-11. It covers God's creation of the universe, the fall into sin, the flood, and the destruction of the Tower of Babel. We do not know the exact date of creation or the time of the flood. We can only start dating events in the world at about 3000 to 4000 years before Christ's birth (B.C.). As a result, we can only make approximations for some of the Bible's earliest dates. We do know that Abraham must have been born near the year 2200 B.C., and we are able to date more accurately events that followed his time.

2) The Patriarchs (2200-1850 B.C.). Genesis 12-50 tells us about this period. The word "patriarch" comes from the Greek words meaning "founding father." It is the word we use to speak about the fathers of the Jewish people, especially Abraham, Isaac, and Jacob. During this period Abraham came to the land of Canaan (Israel). At the end of this period Jacob and his family went to Egypt to live with Joseph because of a famine in the land of Israel.

3) Captivity in Egypt (1850-1450 B.C.). This period is described in the opening chapters of Exodus. The children of Israel were held captive in Egypt for about 400 years. The Bible tells us very little about this period except for the fact that the Egyptians, at some time following Joseph's death, enslaved Jacob's descendants and made their lives very difficult through a program of forced labor. At the end of this period, God called Moses and his brother Aaron to lead God's people out of their bondage in Egypt.

4) The Exodus and the Period of the Judges (1450-1050 B.C.). We learn about this period in the books of Exodus, Leviticus, Numbers, Deuteronomy, Joshua, Judges, Ruth and 1 Samuel. This period starts with the Exodus from Egypt and ends with the anointing of Saul as the first king of Israel. During this period the Israelites wandered for 40 years in the desert after they left Egypt. Also during this period, the children of Israel fought to retake the land of Canaan, and God raised up Joshua and the Judges as leaders of Israel.

5) The United Kingdom (1050-950 B.C.). One can read more about this period in the books of 1-2 Samuel, 1 Kings, and 1-2 Chronicles. This is the time of Israel's first three kings – Saul, David, and Solomon. This period lasted only about 100 years, but it was the time of Israel's greatest earthly rule and accomplishments.

6) The Divided Kingdom (950-600 B.C.). The books of 1-2 Kings and 2 Chronicles and all but the last three of the Old Testament prophets tell of this period. In it, the nation of Israel was divided into two kingdoms who warred against each other – the Kingdom of Israel in the north, and the Kingdom of Judah in the south. The Kingdom of Israel ended in about 725 B.C. when it was overrun by the Assyrians; the Kingdom of Judah lasted until shortly after 600 B.C. when it was destroyed by the Babylonians.

7) The Babylonian Exile and Return to Israel (600-400 B.C.). This period is described by the books of Ezra, Nehemiah, Daniel, Esther and the last three Minor Prophets—Haggai, Zechariah, and Malachi. For 70 years many of the Jews of the Southern Kingdom of Judah were held captive in Babylon. Then the Jews were allowed to return to rebuild the city of Jerusalem and its temple.

8) The Intertestamental Period (400 B.C. – 0). The word "intertestamental" means "between the testaments." This is the name we give to the period of history between the completion of the last book of the Old Testament (the book of Malachi, written about 400 B.C.) and the birth of Christ (about 0). The books which are called the *Apocrypha* (which are not inspired by God) tell us something about the history of this period. During this time, the Greeks, the Jewish family known as the Maccabees, and finally the Romans were ruling over Israel.

9) The New Testament Period (0-A.D. 100). This period covers the time from the birth of Jesus until the death of the last of the apostles, John, who died shortly before A.D. 100. Between these two dates the events of the New Testament took place and the books of the New Testament were written.

These, then, are the nine major periods of Bible history. The dates we have mentioned are not exact but are rounded off to make them easier to remember. Even though we can't always determine exact dates, these will serve as handy references. By learning these periods and their dates you will have a good outline of Bible history to help you in your study of the Bible.

B. MAKING A BIBLE MAP – STEPS 14-16

In the first lessons about making a Bible map we have prepared our paper, drawn the seacoast and edge of the desert, drawn in the Rift Valley and watershed, and added the seven major bodies of water of the region. In step 3, we will add the three major geographical plains of the country. Take a map on which you have completed steps 1 to 13 and follow the next instructions.

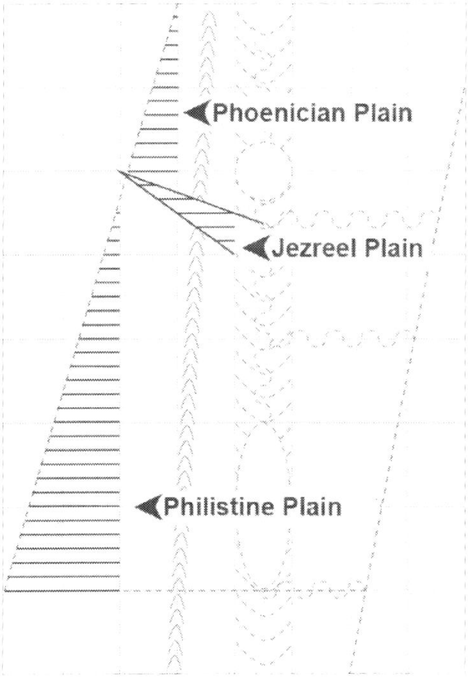

14. Find the point along the seacoast that is even with the bottom of the Sea of Galilee (see the diagram). Draw a series of light horizontal lines that extend from that point to the first fold from the bottom of the map, filling the space between the seacoast and the second fold from the left of the map. This shaded region signifies the Phillistine Plain. This flat area along the seacoast was inhabited by the Philistines for much of the Old Testament period. The Bible refers to it as Philistia (Exodus 15:14; Psalm 60:8; Isaiah 11:14; Amos 6:2). Jeremiah prophesied that the cities of the Philistines would someday mourn when God would bring punishment on them, and there would only be a few people left "on the plain" there (Jeremiah 47:5).

15. Next, draw a series of light horizontal lines in the triangle formed by the seacoast, the third fold from the left and the second fold from the top (see the diagram). This shaded region signifies the Phoenician Plain. The people who lived in this area were called the Phoenicians. They lived in numerous cities and fortresses, the two most famous of which were the important seaports of Tyre and Sidon (2 Samuel 24:7; Isaiah 23:11-12; Joel 3:4). Another important town in Phoenicia was Acco (Judges 1:31), so this area was also at times called the Plain of Acco.

16. Draw a line from the intersection of the second fold from the top and the seacoast to the point where the Jordan River meets the Sea of Galilee. Draw another line from the same starting point to the place where the left edge of the Rift Valley intersects with the third fold from the top. Draw a series of light horizontal lines between these two as the diagram shows. This shaded region signifies the Jezreel Plain which is named after the city of Jezreel towards the eastern end of the plain. Since there were hills to the north and south of this plain, the Bible also refers to it as the Jezreel Valley (Joshua 17:16; Judges 6:33). Since the famous Old Testament city of Megiddo was located near the western tip, that end of the Jezreel Plain was also called the Plain of Megiddo (2 Chronicles 35:22; Zechariah 12:11). The Israelite tribe of Issachar lived in the Jezreel plain (Joshua 19:18). This plain was one of the best areas of Israel for growing crops and thus the prophets used it as a picture of prosperity (Hosea 1:11, 2:22).

These three plains played an important role in the history of Israel. You have now finished steps 1-16. Practice until you can repeat these steps of the Bible map from memory. Your map should look like Map 5.

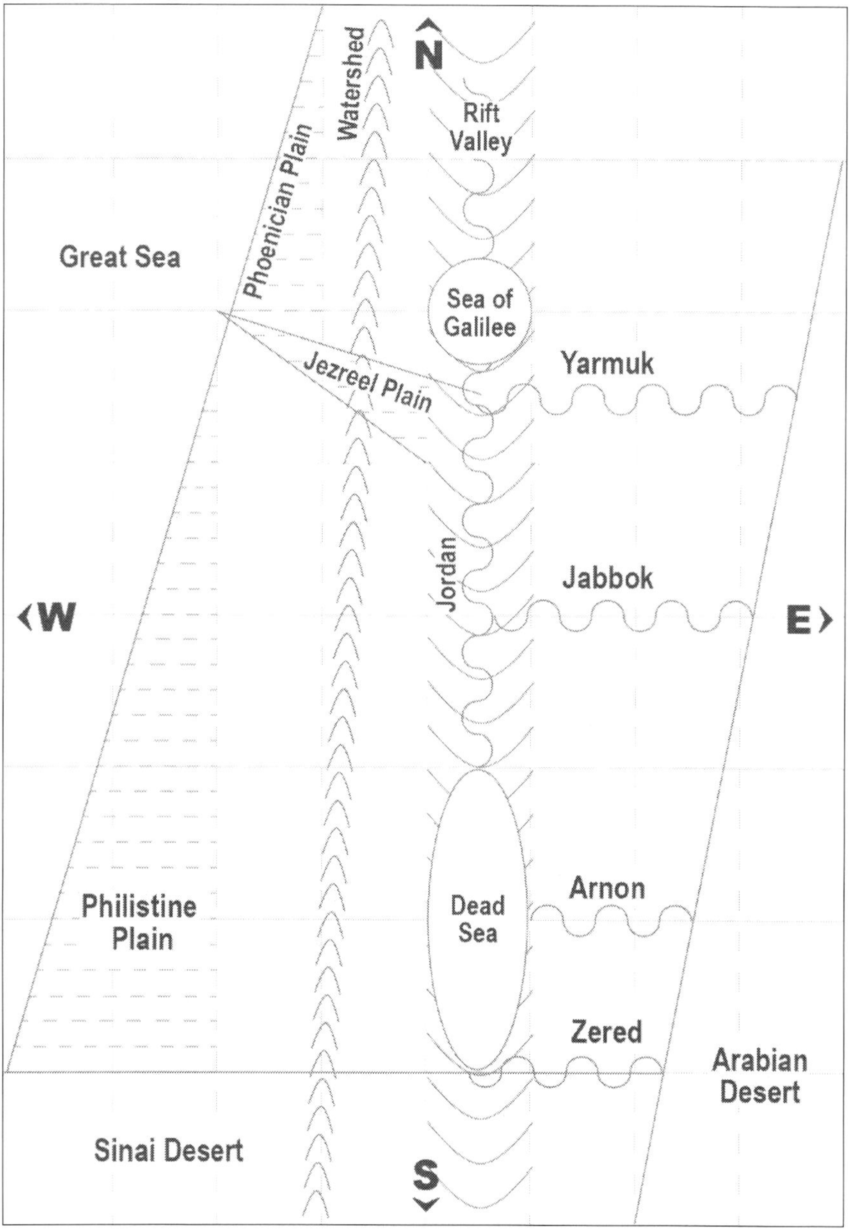

Map 5. The Bible Map –Steps 1-16

LESSON 4:

CIVILIZATIONS OF THE ANCIENT NEAR EAST AND MAKING A BIBLE MAP

(Steps 17-23)

In Lesson 3, we studied the nine important periods of Bible history. We also learned how to put these periods onto a simple time line. In this lesson we will add some more information to that time line. We will add the other important civilizations that existed in the Near East during early Bible times. We will also continue to add additional features to our Bible map.

At the end of this lesson you should be able to:

* *Understand and use the time line of major world civilizations in Chart 2, and*
* *Properly complete steps 17-23 of making your Bible map.*

A. CIVILIZATIONS OF THE ANCIENT NEAR EAST

Look at the timeline shown in Chart 2. The years shown in the timeline—from 3000 B.C. until A.D. 100—are shown at the top and bottom. Right underneath the years at the top, are the nine main periods of Bible history which we studied in the past lesson. On the left side is the word Israel to indicate that these periods designate what was happening in the country of Israel.

Below the word Israel, you see the word Egypt. The information to the right along the time line shows the periods used by scholars to designate Egyptian history. Below Egypt is Syria, then Mesopotamia, and at the bottom is the West, showing what was happening west of Israel, in particular around the Mediterranean Sea. Since we have already studied the periods of Bible history which tell what was happening in Israel, let us begin by looking at the area marked Egypt.

Egypt was the center of a great civilization during most of the biblical period. The Egyptians were one of the first peoples to develop the art of written communication. During the Old Kingdom, the kings of Egypt, called *pharaohs*, built the pyramids as spectacular tombs for themselves.

During the time of the patriarchs in Israel, Egyptian civilization rose to new power in what now is called the Middle Kingdom. Following this period, after several hundred years of fighting and problems, Egypt once again became a world power during the time of the New Kingdom. During the Late Period, Egypt was not as powerful as before, yet it continued to try to influence other countries, especially Israel. Since Egypt was geographically so close to Israel, it is no surprise that from the time of Abraham until the end of the Divided Kingdom in Israel, Egypt continually played an important role in Bible history.

Syria is the area just to the north of Israel. The Phoenicians created one of the first important civilizations in this area. They lived along the sea coast and became great sailors and ship builders. Sailing throughout the Mediterranean Sea, they started many colonies. In the Bible, we often read about the two great Phoenician cities of Tyre and Sidon. The Hittites were another important people who lived in northern Syria. They were a fierce warlike group of people who spoke a language very different from that of their neighbors. Other peoples feared and avoided this nation. During the time of Israel's United and Divided Kingdoms, the Syrian city of Damascus also became the center of an important nation. Sometimes the people of Syria were also referred to as the Arameans.

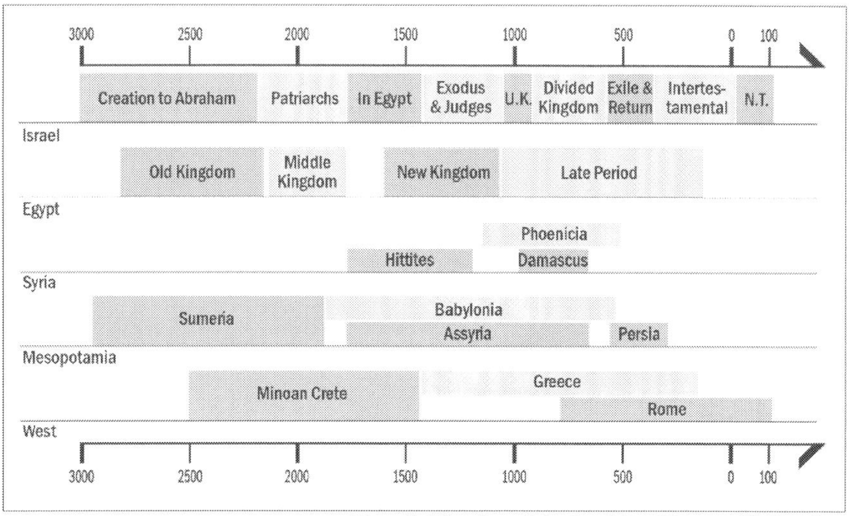

Chart 2. Timeline of Ancient Near Eastern Civilizations

Mesopotamia was the home of many well-known nations. Two of the earliest of these were the Sumerians and Akkadians. These peoples also developed a writing system at a very early period in their history. Sumerian civilization consisted of many city states. One of the most important of these cities was Ur, which might be the Ur where Abraham came from. Later, two other great cities fought to become the supreme ruler of Mesopotamia and the entire Near East. These two cities were Babylon and Asshur. Their civilizations were called Babylonia and Assyria. From the time of the Exodus from Egypt until the time that Judah was carried off into exile in Babylon, these two great powers influenced many of the events in the land of Israel. About 600 B.C., the kingdom of Persia began to develop into a political and military power and eventually replaced Babylon and Assyria as the greatest power in Mesopotamia.

In the west, the first of the great western civilizations arose on the island of Crete. There the Minoans developed an advanced culture that was quite independent of the other civilizations in the Near East. Minoan culture, with its highly developed art, architecture, advanced political structure, and organized trade, became the foundation for the next great western civilization, Greece.

At the same time as the Minoans, an important and influential new culture was in the process of evolving on the Greek mainland and its surrounding islands. Like the Phoenicians before them, the Greeks were great sailors and traders. Over time, they spread their culture throughout the region of the Near East. Much of Greek culture was eventually adopted by the Romans, and from there many ideas and practices were passed on to European and modern Western culture.

The Romans are remembered for building one of the largest and most powerful empires in ancient times. It covered the majority of the ancient Near East, at times extending as far as the Euphrates River. Thus Greece and Rome have had a strong influence on the democratic ideals, codes of laws, art and literary forms of numerous modern cultures.

All of these civilizations have special significance for our study of the Bible and its background. While it is not necessary to memorize all of this information, as we study the different periods of Bible history, we will repeatedly mention these important peoples and their influence. Thus, it is important for students of the Bible to be familiar with them.

B. MAKING A BIBLE MAP – STEPS 17-23

Take a sheet of paper and complete steps 1-16. In this lesson we will add the three large areas to the west of the Rift Valley—Galilee, Samaria, and Judea, and the geographical divisions for each of them.

17. Label the area above the Jezreel Valley and between the Phoenician Plain and the Rift Valley as the Highlands of Galilee. This hilly area to the west of the watershed (along with the Phoenician plain) was given to the tribe of Asher (Joshua 19:24-31), while the hilly area east of the watershed was given to the tribe of Naphtali (Deuteronomy 33:23; Joshua 20:7). The land was good for crops, so it had many storehouses (2 Chronicles 16:4). Later many non-Jews settled in Galilee so it became known as "Galilee of the Gentiles" (Isaiah 9:1). Jesus grew up and did much of his preaching in Galilee. Galileans developed their own accent or dialect, which allowed others to tell where they were from (Mark 14:70; Luke 22:59).

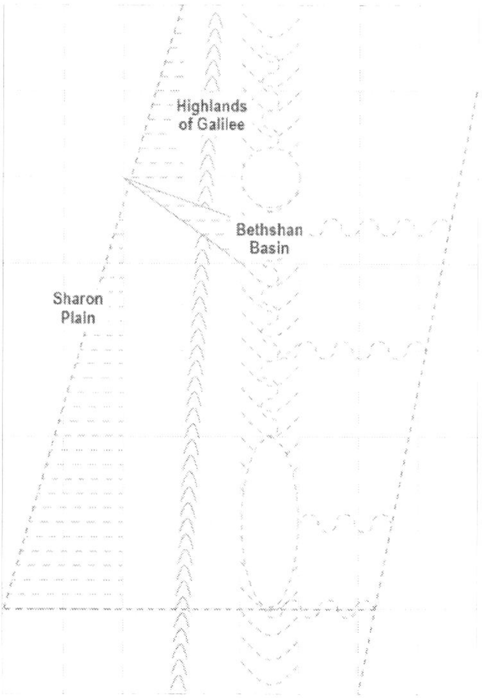

18. Mark the area of the Rift Valley just below the Sea of Galilee as the Beth-shan Basin. This area is named after the city of Beth Shan, where Saul's body was fast-ened after his death (1 Samuel 31:10). Because the Can-aanites had a strong presence in this area, the Israelites, show-ing their weak faith, did not attempt to drive them out and claim the entire land God had promised them.

19. The northern section of the Philistine Plain (starting from the middle horizontal fold on your map) also had a special name – the Sharon Plain.

Mark that on the map. This area was famous as a good place for pasturing animals (1 Chronicles 5:16; 27:25; Isaiah 65:10), and for the beautiful flowers that grew there (Song of Solomon 2:1). Thus it became another symbol for prosperity (Isaiah 35:2), and Isaiah pictured a time of great trouble as a time when the Sharon Plain would become as dry and desolate as the Arabah around the Dead Sea (33:9).

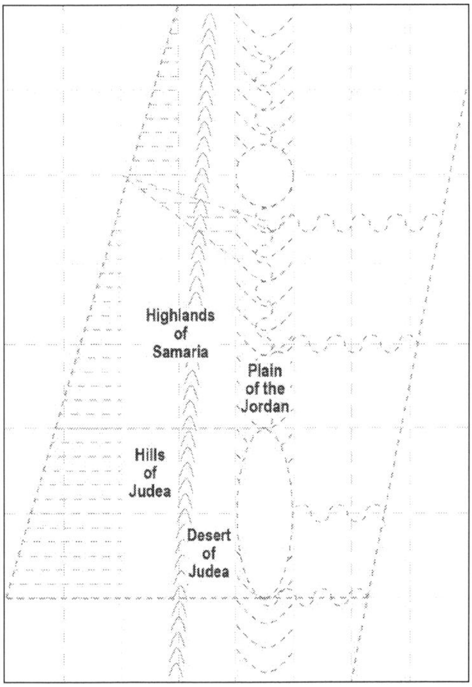

20. Draw a dotted line from the top of the Dead Sea straight west to the seacoast. Label the area above that line as the Highlands of Samaria (Amos 3:9). Samaria began as the name of a single hill in this area, but the Israelite king Omri built his capital city there and called the new city Samaria (1 Kings 16:24). Soon the entire area was called Samaria. Since this area was given to the tribe of Ephraim, it was also called the hills or hill country of Ephraim (Judges 3:27; 2 Chronicles 15:8; Jeremiah 4:15). When the Assyrians carried many Jews among the northern tribes of Israel into captivity, they brought in new people to live in this area, people who worshiped different gods (2 Kings 17:24-25, 29). From then on the Israelites viewed the people of Samaria as their enemies. Even in New Testament times the Jews often only traveled through Samaria or had dealings with the people who lived there (known as Samaritans) when they were forced to (John 4:9). The area, however, was very fertile and good for growing crops (Jeremiah 31:5).

21. Mark the part of the Rift Valley just above the Dead Sea as the Plain of the Jordan. The Hebrew Old Testament uses the word *kikkar,* meaning "circle," to describe this area. This was an area that was well-watered by the

Jordan River, so many trees, plants, and wild animals were to be found there (Zechariah 11:3). It was also here that the children of Israel crossed over the Jordan River.

22. Mark the area between the watershed and the Dead Sea as the wilderness or Desert of Judea. This barren area was where John the Baptist preached (Matthew 3:1) and where Jesus fasted for forty days and nights (4:1). Food and water were scarce there (John had to eat grasshoppers and wild honey [3:4]), and few living creatures, except snakes (3:7), could survive the harsh environment. It was not a place where people normally lived or even visited, so it was amazing that so many went there to hear John's preaching (3:5).

23. Mark the area between the watershed and the Philistine Plain as the Hills of Judea, the name for this area in New Testament times. In the Old Testament it was called the hills of Judah (2 Chronicles 21:11). This was the prime area in the south for farming and pasturing (Joel 3:18).

This completes steps 17-23 of making your Bible map. Your map should now look like Map 6.

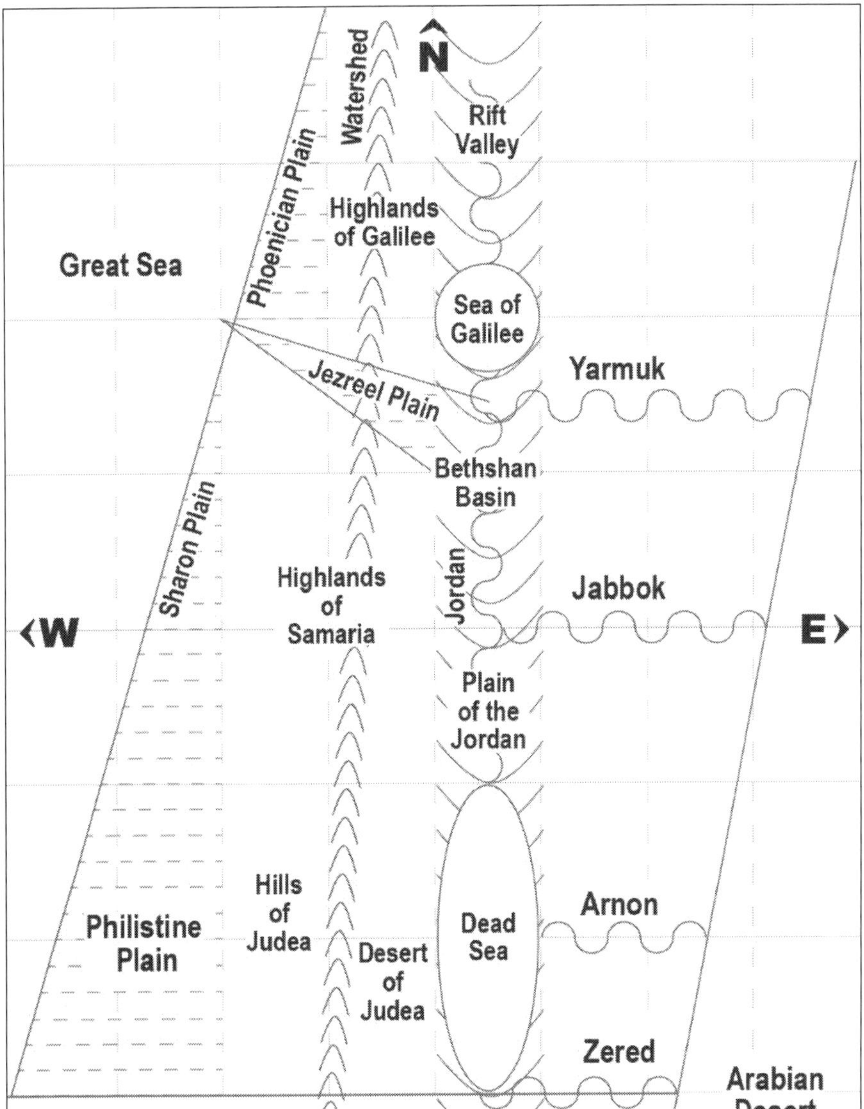

Map 6. The Bible Map – Steps 1-23

LESSON 5: BIBLE MAP
(STEPS 24-27)
AND USING THE BIBLE MAP

In our first lessons, we began drawing a Bible map. In this lesson, we will complete our map and also learn how we can use it in our study of the Bible.

At the end of this lesson you should be able to:
* *Properly draw the map of the Holy Land; and*
* *Label the map with the 26 proper names we have learned.*

A. MAKING A BIBLE MAP – STEPS 24-27

To complete our Bible map, we only need to add four more areas. The first three of these areas are border areas. They lie between the major political areas and served as buffer areas. The control and loyalty of these areas often switched from one of its neighbors to the other.

24. Mark the eastern edge of the Philistine Plain next to the Hills of Judea as the Western Foothills (see the diagram). The Hebrew word for this area is *shephelah.* It designates an area of low hills between the coastal plain and the higher hills of Judea (Deuteronomy 1:7; Joshua 9:1; Judges 1:9; 1 Chronicles 27:28; Jeremiah 17:26; Zechariah 7:7). It is sometimes also referred to simply as "the foothills" (1 Kings 10:27; Obadiah 1:19). The Israelites often fought with the Philistines for control of this area.

25. Label the southernmost area of Samaria, from the seacoast to the Rift Valley, as the Corridor. It is not mentioned by name in the Bible, but this strip of land served as a buffer area between Samaria and Judah.

26. Label the area of Judea even with the bottom of the Dead Sea as the Negev. It was the border area between Judea and the Sinai Desert. This area was semi-desert, but herds could graze there part of the year, and wild animals lived there as well (Isaiah 30:6). That is why Abraham and his

descendants spent time there (Genesis 12:9) but could not stay too long in any one place (13:3). During the hotter and drier seasons, shepherds moved their herds into the hills of Judea to find food (Numbers 13:17). The only people who could live in the Negev the entire year were those who lived by one of its springs (Joshua 15:19). The tribe of Judah conquered part of the Negev and lived there, but other non-Jewish tribes continued to live there as well (1 Samuel 27:10). The Bible pictures God's unexpected blessings to be like the streams of water which suddenly appeared in the Negev after the very rare thunderstorms (Psalm 126:4).

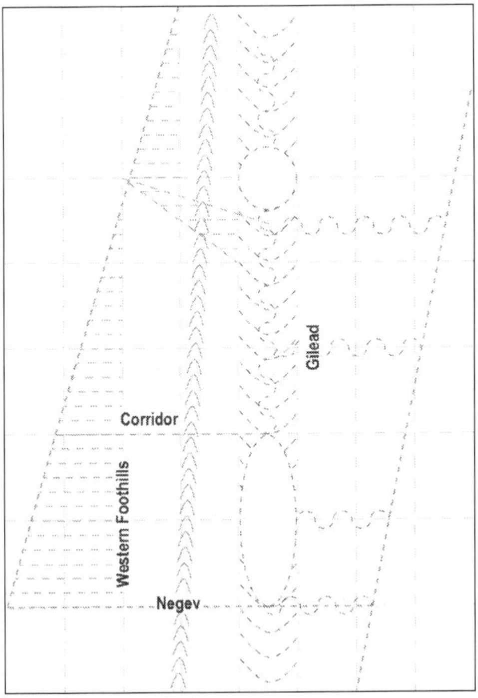

27. Find the area across the Jordan River from Samaria. Label this area just east of the Rift Valley, extending on both sides of the Jabbok River, as Gilead. Gilead was a hilly region (Genesis 31:21) that was good for raising livestock (Numbers 32:1). God gave this region to the Israelite tribes of Gad and Reuben (1 Chronicles 5:1, 9), but part of the tribe of Manasseh lived there as well (Joshua 17:6; 22:13). Gilead also became famous for the balm or medicine produced there (Jeremiah 8:22; 46:11).

You have now completed the Bible map and have marked 26 important features on it. In many coming lessons, we will make use of this map. As we study the different periods of Bible history we will be placing the important towns and countries of each period on the map. Be sure that you have now mastered the steps. Practice them until you can make and label the map perfectly every time. You will find that your knowledge of the Bible map will be useful for years to come as you study your Bible. Your completed map should look like Map 7.

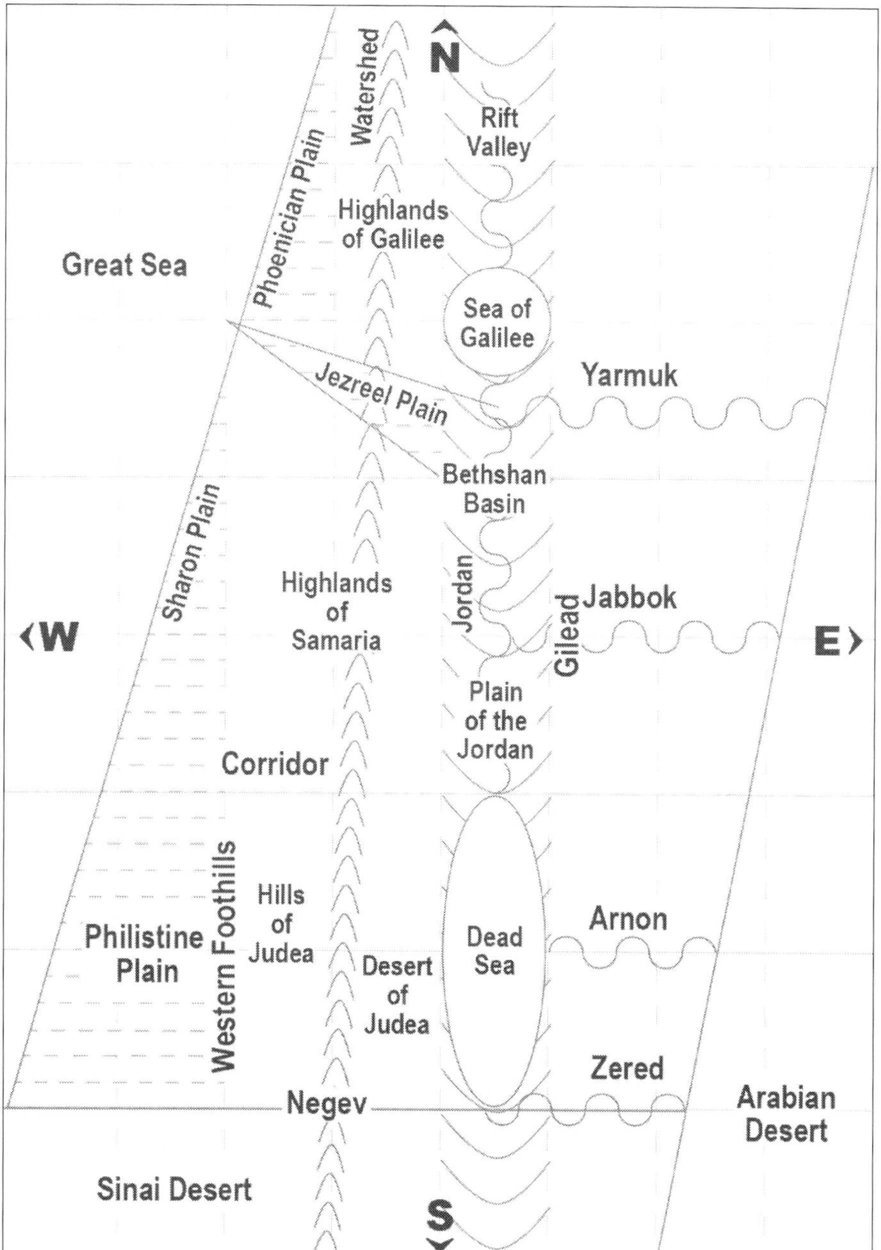

Map 7. The Completed Bible Map

B. USING THE BIBLE MAP

Put your completed Bible map in front of you and we will look at the different areas shown. Notice how the land of Israel is protected on three sides. On the east is the Arabian Desert; to the south is the Sinai Desert; and to the west is the Mediterranean Sea. Because the deserts and the sea protect Israel on these three sides, most of Israel's invaders came from the north. Thus the prophets often spoke of God's judgment coming from the north in the form of invaders (Isaiah 14:31, 41:25; Jeremiah 1:13-14, 6:22). Occasionally, however, the Egyptians crossed the Sinai Desert and invaded from the south.

Now look at the Rift Valley. Notice how it divides the land into two parts. The area to the west of the Rift Valley is today called the West Bank. In both ancient and modern times, the West Bank was the heartland of the people of Israel. The area to the east of the Rift Valley is called the East Bank. It is sometimes also referred to as the Transjordan, which comes from the Bible expression "across the Jordan" (Matthew 4:25; Mark 3:8, 10:1).

Let us look next at the East Bank. Two and a half of the twelve tribes of Israel lived in the area known as Gilead. Also living on the east side of the Jordan were Israel's cousins: Edom lived in the area below the Zered River and in the Rift Valley south of the Dead Sea; Moab lived along the east side of the Dead Sea, on both sides of the Arnon River; and Ammon lived in the area between Gilead and the edge of the desert. To the north of the Yarmuk River was the land of Bashan, famous for its large bulls (Psalm 22:12). All of these lands stretched from the east side of the Rift Valley all the way to the Arabian Desert, and they lay on plains or hills high above the deep valley.

Separating the East Bank and the West Bank is the Rift Valley. Within this valley, from north to south, are: 1) the sources of the Jordan; they flow into 2) the Sea of Galilee; 3) the Bethshan Basin below the Sea of Galilee; 4) the Jordan; it widens out into 5) the Plain of the Jordan just above the Dead Sea; and finally 6) the Dead Sea itself. Note that the Dead Sea has no outlet. The water and its minerals stay in the sea or are evaporated by the heat. The Rift Valley is truly a deep valley. In fact, on all the seven continents, the Dead Sea area is the lowest spot below sea level on any continent.

Look now at the West Bank. Note that it is divided into three main areas (from north to south) – Galilee, Samaria, and Judea (or Judah). The other nine and a half tribes of Israel lived in the West Bank. The Jezreel Plain separated Galilee from Samaria. The corridor separated Samaria from Judea. The Negev separated Judea from the Sinai Desert. In Galilee, starting at the seacoast, the ground rises quite steeply from the Phoenician Plain into the highlands of Galilee and, after crossing the watershed, falls again to the Rift Valley. In Samaria, the Philistine and Sharon Plains lie along the coastline and the land to the east rises more gradually to the watershed. From there it descends more gradually down to the Rift Valley. In Judea, the Philistine Plain leads into the western foothills, and then into the hills of Judea; after the watershed the Desert of Judea leads down to the Dead Sea.

Now look at the seacoast. Notice that it is very straight and therefore had no good harbors. Because of this, Israel never became an important sea trading nation. The one good harbor was at Acco (Judges 1:31), called Ptolemais in New Testament times (Acts 21:7). It was located just above where the Jezreel Plain meets the seacoast (see Map 17). Along the coast are the broad Philistine, Sharon, and Phoenician Plains. This made Israel somewhat vulnerable from the sea. However, further inland, the western foothills and hills of Judea and the highlands of Samaria and Galilee provided a natural protection from invaders. A few important valleys led through these hills up into central Israel, so the Israelites built fortresses and cities above these valleys to protect themselves from invaders.

Now that you have learned to draw this map, you will be able to understand many Bible passages much better. For example, read Moses's words to Israel in Deuteronomy 1:7:

> Break camp and advance into the hill country of the Amorites; go to all the neighboring peoples in the Arabah, in the mountains, in the western foothills, in the Negev and along the coast, to the land of the Canaanites and to Lebanon, as far as the great river, the Euphrates.

You can now understand that Moses is telling the Israelites to go up among the people living in the Arabah (here specifically the area of the Rift Valley below the Dead Sea), in the mountains or hills of Judea, in the western foothills, and in the Negev below Judea, and all the way up the seacoast to Lebanon (Syria), off the top of our map.

Or if you read in Judges 1:9 that "the men of Judah went down to fight against the Canaanites living in the hill country, the Negev and the western foothills," you can picture in your mind that the men of Judah were fighting their neighbors just to their west who were living in those buffer areas between themselves and the seacoast—in the hills west of the watershed, the western foothills and the Negev.

As we continue our study of various periods of Bible history, we will refer to our map many times. However, if you have learned to make the map, you should already be familiar with the most important geographical and political features of Israel.

LESSON 6: FROM CREATION TO ABRAHAM

As we continue our study of Bible background, we will briefly examine the nine main periods of Bible history. As we study each period, we will also learn about the other important civilizations of the Middle East that existed during the same period. We will further note the customs and living conditions of these peoples. In this lesson, we will begin with the first period of biblical history – from creation to the time of Abraham.

At the end of this lesson you should be able to:

* *List the five most important points of the Bible's account of creation; Describe the teachings of evolution and point out its fallacies;*

* *Give two examples of why we can say that the civilization between creation and the flood was an advanced civilization;*

* *Name the four major facets of the Bible's story of the flood;*

* *Explain how God's reason for sending the flood differs in the Bible from the Mesopotamian story about a great flood; and*

* *Tell what the Table of Nations is and where it is found.*

A. CREATION

The Bible begins with God's creation of the universe. The Bible does not provide us with the exact date when this happened; it only states that it occurred "in the beginning." Father, Son, and Holy Spirit were all present and each participated in the act of creation. Up to that point nothing existed. We therefore have no way of knowing how long ago the universe was created. We might guess that it took place sometime between 5000-10,000 B.C.

The actual date of creation is not in itself important. What is important are the following five points:

1) God created all things out of nothing through the power of his Word.

2) God created man as a special being – perfect and without sin.

3) God created woman to be a fitting companion for man.

4) Adam and Eve fell into sin, thus bringing death into the world.
5) God immediately promised to send a Savior from sin.

Most of the world's tribes and peoples have their own stories about how the world and mankind came into being. Most of these stories acknowledge that some kind of supreme being created the world. However, almost all these stories place some limits on that creator's power or wisdom. Some of the stories even have striking similarities with the Bible's account of creation. None of them, however, fully agree with the five main truths about the creation story in God's Word. It may be that these stories were originally correct. But after being handed down from person to person over hundreds of years, they were gradually changed, and false ideas were added. Because of this they no longer relate the true story of creation. Only the book of Genesis has preserved the truth about creation.

The civilizations in the ancient Near East also had stories about the world's creation. Since these stories are very old, some people have claimed they give a more ancient account about creation. Others say these stories may be the source behind the Bible's account. For instance, a very old Mesopotamian story tells of a battle between the gods at the time of creation. One of the gods was supposedly killed, and his body was used to make the earth. The gods then created men to work for them as slaves. It is not difficult to see why we do not accept this and similar accounts. They do not agree with the truthful account God has given us in the Holy Scriptures.

Many modern scholars and scientists also attack the biblical account of creation. They refuse to accept that the world could have been created in only seven days. Instead, they accept the theory of evolution which teaches that the world came about naturally without the influence of God. This, in turn, leads them to deny the entire Bible story or to alter it to fit these new theories. Christian scientists can point out various flaws in these other theories and find evidence to support the Bible's account of creation. While this evidence is reassuring, we reject evolution and believe the Bible's account of creation simply because God tells us so. Thus, we believe that in six normal days (Genesis 1 mentions *evening* and *morning* for each day) God created our universe and everything in it. We also accept the fact that the world was perfect in the beginning, that man sinned and thus brought death into the world and that in order to be restored to life, mankind needs a

Savior. We believe all this to be true because God has proclaimed these things as facts in his inerrant and infallible Word.

B. BETWEEN CREATION AND THE FLOOR

Genesis chapters 4 and 5 give us our only information about the time between creation and the flood. These chapters begin with the story of Cain and Abel and then list the names and ages of various descendants of Adam. The Bible states that some people were city or urban dwellers (4:17), while others lived in rural settings (4:20). Some, like Cain, became farmers; others, like Abel, had flocks of animals. People learned to play musical instruments (4:21) and made bronze and iron tools (4:22). The faithful also organized worship services and preached about God (4:2).

As mentioned before, we do not know for sure how many years passed between creation and the flood. In Genesis 5, the ages of many individuals are given, but we have no way of knowing if God listed all the names or just the important ones. From the ages given we can figure that at least 1650 years passed between creation and the flood; but depending on how many names were not included, this time may have been considerably longer. By the end of this period, the people had become extremely wicked. They no longer worshiped the true God or followed his commandments. As a result, God ended this period of history with the great flood.

C. THE FLOOD

The world's earliest civilizations, which developed between the time of creation and the flood, ended in complete failure. They had achieved notable accomplishments working with different types of metals, composing and playing music, and building and living in cities. Yet they had neglected what was most important—their relationship with God. As their wickedness increased and they fell further and further away from God, God decided to destroy them with a universal flood. First, however, God in his grace and mercy sent Noah, a preacher of righteousness (2 Peter 2:4-9), to urge these wicked people to repent of their evil ways. Their rejection of God's grace led to their utter destruction.

The Bible tells us the following important facts about the flood:

1. The flood was universal (it covered the entire earth).

2. Only eight people, Noah and his relatives, were saved because they believed what God had revealed to them.
3. God sent the flood as a punishment for mankind's sin.
4. Except for the animals able to live in water and the people and animals which Noah took with him in the ark, all living creatures perished.
5. The flood lasted just over a year and ended when God allowed Noah, his family, and the animals on the ark to begin life again on dry land.

The story of a great flood is found in some form in almost every tribe and culture throughout the world. Many of these stories are hundreds or even thousands of years old and may be stories about the real flood. However, these no longer tell the story accurately. Just as with human accounts about creation, so also with these variants on the story of the flood, we need to remember that only the Scriptures give us a fully reliable account.

A good example of a distorted flood story is one from Mesopotamia. It has many similarities to what the Bible teaches, but also a number of important differences. It begins by relating that the gods became angry with mankind. Men were making so much noise that the gods were unable to sleep properly. In their anger, the gods decided to destroy mankind with a great flood. One of the gods, however, warned a man named Atrahasis. Like Noah, Atrahasis built a boat, placed some animals inside, and was saved from the devastation of the flood. This account also states that the gods began to suffer because they had no food during the period of the flood since all the animals had been destroyed. So after the flood ended, the gods swore they would never again destroy the world by means of a flood. This account even mentions that Atrahasis's boat landed on a mountain, and that he had sent out birds to see if the land had dried out.

While the Mesopotamian flood story has many similarities to the story of the flood in Genesis 6-9, it also has significant differences from God's revelation about this event. Note especially how point number 3 above has been changed. In the Mesopotamian story of the flood, the gods destroy mankind because people were making too much noise. Would a just and gracious God even consider such a thing? The Bible clearly explains that the Lord wiped away mankind and began anew at that time because of mankind's total sinfulness and refusal to listen to God's message and repent. God's punishment was just, and he displayed his undeserved love by allowing mankind to begin once again through Noah. Again, what the Bible teaches gives us the only true account of what took place at the time of the flood. It also shows us clearly what our God is like and why he acts as he does.

D. THE PERIOD FOLLOWING THE FLOOD

After the flood, the world was soon filled with the descendants of Noah. Genesis 10 and 11 recount this period for us. Chapter 10 provides a list of nations which came from the sons of Noah. This list is commonly referred to as the Table of Nations. We can no longer identify where all these nations were located. We can only place some of them. For more information about the Table of Nations, consult a study Bible or Bible dictionary. Once again, note that Israel is located at the center of all these countries of the Near East.

It is also worth noting that although there were many nations in the world at this time, they all spoke the same language. We are not certain what that language was. However, we do know from what God has told us in Genesis 11 that it was mankind's pride and arrogance which led them to reject God's command to spread out over the face of the earth (Genesis 9:7, 11:4-9). This sinful pride resulted in the rebellious building of the tower of Babel. This, in turn, led God to confuse the people by giving them different languages. Since they were unable to speak or work together effectively, each group separated and spread out to live by itself. In this way, God both placed a check on the world's rebellious spirit and ensured that his plan for populating the entire earth would take place. New civilizations began to develop, but sadly, many of the people of the world continued to go their own sinful ways. Our next lesson will focus on the development of one of those early civilizations which emerged after the flood – the Sumerians.

LESSON 7:
THE SUMERIANS

In our earlier lessons we noted that many of the ancient world's most important civilizations arose along the Fertile Crescent. In this lesson we will study the earliest of all Mesopotamian civilizations—the Sumerians.

By the end of this lesson you should be able to:
* *Name the two areas of Mesopotamia where the Sumerians lived;*
* *List the four most important Sumerian cities;*
* *Give the Sumerian word for king and the name for Sumerian writing; and*
* *Identify the early famous king of Uruk, the first emperor of Sumeria, and the most famous citizen of the city of Ur.*

A. SUMER AND AKKAD

In Lesson 6 we made reference to the Table of Nations which lists the descendants of Noah and his three sons. Among those descendants was a man named Nimrod whom the Bible describes as "a mighty warrior on the earth" (Genesis 10:8). From this and the verses following, we are able to infer that Nimrod founded a mighty kingdom whose first centers were Babylon, Uruk (Erech in some Bible translations), and Akkad, all located on the plain of Shinar. Another name for this region is Sumer. Sumer was located on the large plain between the Tigris and Euphrates rivers, at the southeast end of the Fertile Crescent. A bit farther to the northwest, also between these two rivers, was the area called Akkad. It was on these two plains, both in what is now the country of Iraq, that the first great civilization of Mesopotamia arose. Although the Sumerians and Akkadians were distinct peoples, we will consider them together and refer to them simply as the Sumerians.

The original settlers of this region were mainly farmers. They made use of the good land and the abundant supply of water to grow their crops. As time passed and threats were made on their livelihood, the people began living together in cities for protection. By about 3000 B.C., the Sumerians had begun building cities. Around this same time they began to divide their labor – some farmed, others cared for animals, and still others served as soldiers and hunters. From these early roots came the foundation of Sumerian civilization.

B. SUMERIAN CITIES

Cities were the center of Sumerian culture. The Sumerians built many large, organized and important cities. Since each functioned independently, we should think of them as city-states. Some of the most important were Ur, Lagash, Uruk, and Babylon. These early cities were completely surrounded by mud-brick walls to protect them from wild animals and human enemies. Around each city were fields of crops irrigated by a system of canals which improved their harvests. In each city were temples dedicated to various gods and goddesses. They had full-time priests to offer up sacrifices and prayers and to take care of the temples. Each city had its own guards to offer protection and an army to wage war against enemies. The very wealthy could afford their own large homes, but the majority of people were poor and lived in simple one-room huts. Uruk was especially prosperous in early times. By 3000 B.C. it was the largest city in the world, and its walls enclosed an area as large as that of Jerusalem in the time of Jesus.

At first, citizens met together to make important decisions and ruled over each city. Before long, however, governors took control of each city. Eventually one governor took complete power and became a supreme ruler or king. The Sumerian name for king was *lugal*. The Sumerians came to believe that the gods had sent the lugal to be their ruler. Along with the priests, the lugal exercised absolute control over the city. Everyone else was required to pay taxes to support both the king and those who served at the temple.

C. THE SUMERIANS DEVELOP THE ART OF WRITING

Sometime between 4000-3000 B.C. the Sumerians developed a system of writing. They wrote on small tablets made of soft clay (the invention of paper lay well in the future). A sharpened reed was used to make their characters in the soft clay. If a document needed to be preserved, they simply baked the clay tablets. Today thousands of these clay tablets have been discovered, studied, and translated by archaeologists and scholars. They give us important glimpses into ancient Sumerian life and society.

At first the Sumerians merely used pictures to convey what they wanted to express. Soon they developed symbols to represent different sounds and syllables. Hundreds of different symbols were needed to write their language. Obviously, this made writing very complicated and cumbersome. Only specially trained people, such as secretaries and priests, were able to read and write. Because of the wedge-like shape of the lines made by their reed pens, this type of writing has been given the name cuneiform.

The language of the Sumerians seems unrelated to any other language. The language of the Akkadians, however, is a Semitic language (related to Hebrew) and is one that we can decipher. Yet both of these languages were written using similar cuneiform symbols. Initially, most writing was for receipts and other business purposes. Soon, however, priests used writing to record stories about the gods. In our last lesson we discussed some of the stories regarding creation and the flood that these early Mesopotamians had written. Records of their laws, hymns, and prayers to their gods, as well as tales about their great kings and heroes, have also survived. Thanks to the survival of their ancient clay tablets, we now know a great deal about the history of these early peoples.

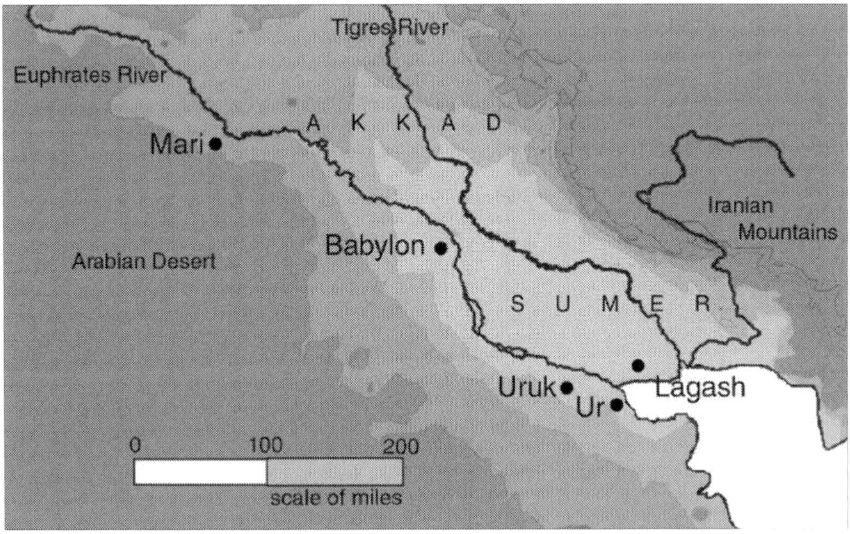

Map 8. Sumeria

D. SUMERIAN RELIGION

The Sumerians were polytheists – they had many gods, both male and female. They created a god or goddess for almost every area of life. An, for example, was the god of the sky; Enki, the god of water; Ninkhursag, the goddess of the earth; and Ishtar, the goddess of love and fertility. Most other civilizations of the Near East also had this idea of a multitude of different gods who watched over the different areas of people's lives. Neighboring peoples also worshiped many of the Sumerian gods. During the time of Ezekiel, for example, some elders of Israel were found worshiping Tammuz, the Sumerian god who was thought to be the husband of Ishtar (Ezekiel 8:12-18).

Each city had a god or goddess who was believed to be the special protector for that particular city. The city's inhabitants gave these deities special worship and sacrifice in order to keep them happy or in order to gain some special help. When one city conquered another, it was thought that the god of the victorious city had defeated the god of the other city. The Sumerians made statues of their gods and built temples to house them as part of their worship. Religious life was a vital element in the life of every Sumerian, and it played an integral role in the development of their civilization. Sumerian gods were like human beings in many ways. Although far more powerful than ordinary humans, these gods were thought to have human emotions like love and joy and to take part in sinful behavior such as lust, jealousy, and deception. As with so many other ancient civilizations, the Sumerians sought to create god in their own image. These peoples had quickly forgotten the truth about God's creation, no longer remembered God's warning in the flood, and had arrived at the point where they lost or ignored their knowledge of the true God! As has so often happened in history, they forgot the creator and worshiped his creation instead.

E. FAMOUS SUMERIANS

Earlier we learned about the famous warrior Nimrod who was one of the founders of the early Sumerian kingdoms. Nimrod was just one of a long line of famous Sumerians. According to legend, Gilgamesh was an early king of Uruk who was remembered for his mighty deeds. Another famous ruler was Sargon I (2275-2220 B.C.), a king of Akkad. He conquered most of Sumer and became what might be called the first emperor of the Middle East. Sargon's grandson Naramsin also ruled all of

Akkad and Sumer. He took for himself the title "King of the Four Quarters of the World" although he was only king of the Sumerians.

Other famous rulers came from the well-known city of Ur, which dates back at least to 2500 B.C. From 2150 until 1950 B.C. Ur was the dominant city within Sumeria. The great king Ur-Nammu brought the city to power, and his successors maintained it. The most famous citizen of Ur, however, was not a king or even a priest. He did not worship the Sumerian gods or follow the Sumerian way of life. Rather he worshiped and revered the one true God, the LORD Almighty. At God's command he packed up his belongings and, together with his household, left the mighty city of Ur in order to live in a foreign land. This descendant of Noah's son Shem was the patriarch Abraham. Abraham became the father of many nations, and through his descendants the Savior came into the world.

LESSON 8:
THE LIFE OF THE PATRIARCHS

In Genesis 12-50 we are given the history of Abraham, Isaac, Jacob, and Jacob's 12 sons (who later on became the basis for the 12 tribes of Israel). These founders of the nation of Israel are referred to as the patriarchs. We will not review their lives in our course, but we will discuss *how and where* they lived.

At the end of this lesson you should be able to:
* *Name the five kinds of animals that the patriarchs owned;*
* *Give at least two names for people who move from place to place;*
* *Describe what a tent was like in the time of the patriarchs;*
* *Name the three areas of Israel where the patriarchs lived most of the time;*
* *Describe the typical clothing of men and women at Abraham's time; and*
* *Describe the common food of the patriarchs.*

When Scripture references are given in this chapter they will be identified by book, chapter and verse, i.e., Isaiah 55:12. When no Bible book is given, the reference is to the book of Genesis from which most of the references in this lesson are taken.

A. NOMADIC LIFE

The patriarchs, under God's blessing, were rich people. They owned gold and silver, possessed numerous servants (14:14 and 17:27), and had other possessions as well. Their greatest source of wealth, however, was their animals. They had large herds of sheep, goats, cattle, donkeys, and camels (24:10). Even without their other possessions, just owning these large herds made them very rich. These herds also affected the lifestyle of the patriarchs. They had to spend much of their time moving from place to place in order to find sufficient pasture for their flocks.

People who move from place to place with their animals are called nomads or Bedouin. Today there are still people in the Middle East who live their life as nomads or Bedouin. Until recently, they lived very similarly to how nomads lived in the time of the patriarchs.

Abraham and his descendants were only partly nomadic. While they often moved from place to place, at other times they settled for longer periods of time and even grew crops (26:12). Because of their wealth, some of the patriarchs were even able to settle down within urban areas while their servants and other members of their family took care of their herds (20:1). For much of the time, however, the patriarchs lived in tents. The typical tent was made from the hair of goats or camels. Such goat-hair tents would let the breeze pass through on hot days, yet they remained waterproof when it rained. These tents were usually quite large and were divided into sections to form separate areas for men and women, as well as a place to receive visitors.

The tents were held up by poles with ropes tied to stakes to hold them down (Judges 4:21). Curtains hung down at the front of the tent for privacy and protection from the sun, dust, wind and rain. As a family grew, the tent could be made larger. God used this picture of enlarging one's tent in Isaiah 54:2-3 as a beautiful picture of the happiness and success that God would grant to his church in the New Testament period.

Inside most tents there was little furniture. People did not use chairs but rather sat on mats or the saddles from their camels (31:34). Since everything had to be moved quickly and easily as the family followed the herds, large pieces of furniture could not be used. Mats or rugs covered the floors and served as beds as well as chairs. In the cold and rainy season, a small burner placed inside the tent provided some warmth. Valuables were buried somewhere under the tent for safe keeping. When it was time to find new pasture, the tent and all its belongings were taken down, folded up, and loaded onto camels or donkeys.

B. WHERE THE PARTIARCHS TRAVELED

Look at Map 9. The map shows nine cities that were important during the time of the patriarchs. Notice that five of the cities are along the same line—the watershed line. As you learned in Lesson 2, this line runs along the highest hills that form a ridge down the center of Israel. All the rivers east of this line flow into the Rift Valley, and the few streams on the west run into the Mediterranean Sea. The patriarchs spent much time along the watershed because this land was not occupied by settlers and was therefore more free for grazing their large herds. Also, during the hottest times of the year they could still find good pasture for their animals there.

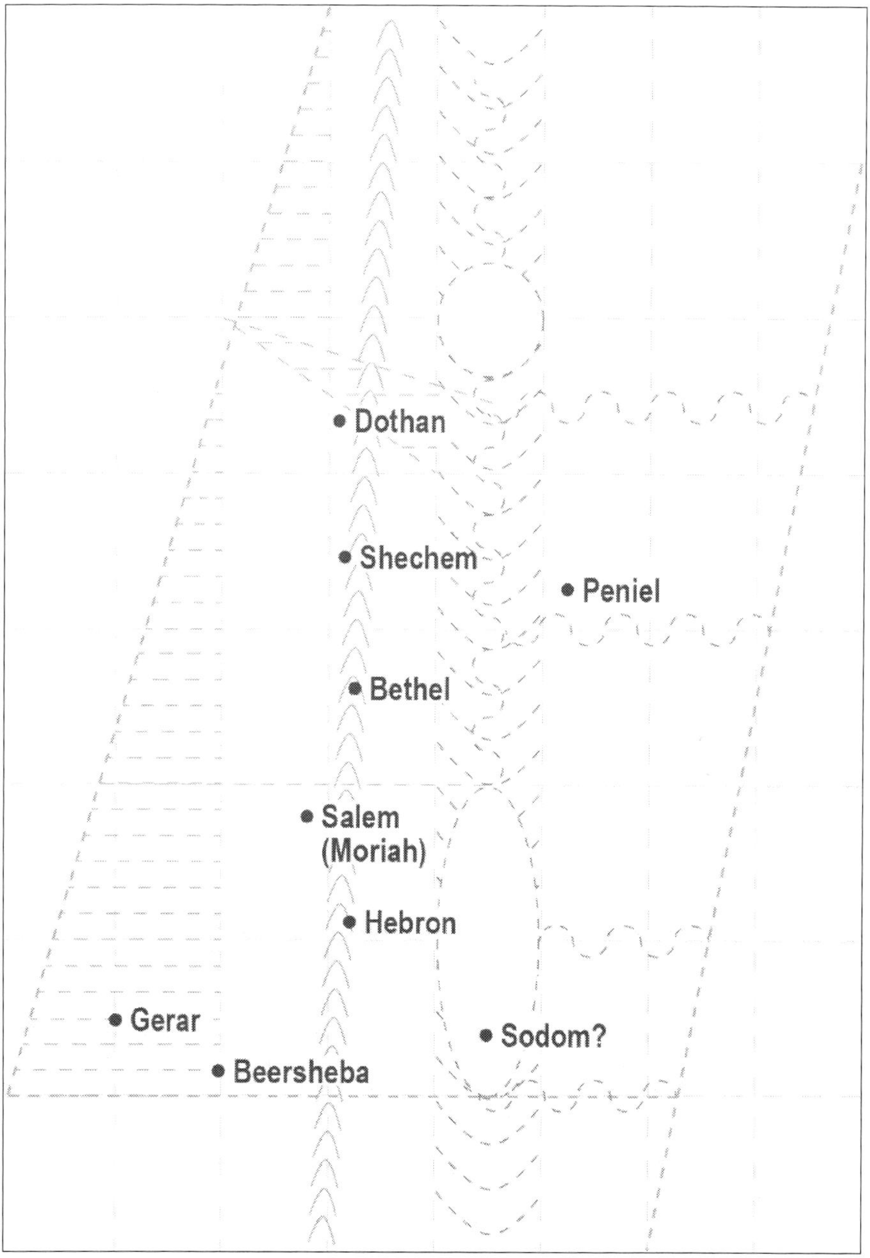

Map 9. Israel in the Time of the Patriarchs

The three main areas where the patriarchs lived were Bethel, Hebron, and Beersheba. When Abraham first arrived in the land of Canaan, he lived for a while at Shechem (12:6). His grandson Jacob even purchased some land there (33:18-20). Not far from Shechem, Jacob wrestled with the Lord at Peniel along the Jabbok River as he returned again to his homeland. Jacob settled more permanently at Bethel (35:1 ff.), where earlier he had his famous dream while fleeing from his angry brother Esau. Salem (later called Jerusalem) was where the priest Melchizedek served as king. He received Abraham's gifts following Abraham's rescue of Lot, (14:18 ff.). The highest point of Salem was Mt. Moriah, where God had asked Abraham to sacrifice his son Isaac (22:1ff.). Gerar was the city where Abimelech was king and where Abraham stayed several times (20:1ff.), as did his son Isaac (26:1). The cities of Sodom and Gomorrah, destroyed by God for their wickedness (19:24ff.), were probably located somewhere under or just south of the Dead Sea. Dothan was where Jacob's brothers sold him into slavery (37:17). Hebron, with its cave of Machpelah, is the burial site of Abraham, Isaac, and Jacob (25:9-10, 35:27-29, 50:12-13).

C. DAILY LIFE

Much about everyday life (food, clothing, entertainment, and family life) remained unchanged throughout the Old and New Testament periods. For example, the clothing worn by the patriarchs was quite simple, and it remained quite similar to that worn by Jesus's disciples two thousand years later. A loose-fitting cloth around their waist served as an undergarment (Exodus 28:42). A long loose robe which extended from the shoulders to the ankles served as an outer garment (1 Samuel 2:19). Usually this was plain, but at times it could be dyed with beautiful colors or decorated (Genesis 37:3; 2 Samuel 13:19). A belt or girdle secured this garment about the waist (2 Samuel 20:8; 2 Kings 1:8). When running or hurrying, a person often tucked a fold of his robe into the belt to shorten the robe and keep it above his knees (Exodus 12:11). Sandals were worn to protect the feet (Deuteronomy 29:5), and a cloth was usually tied around the head for protection from sun, wind, and rain. In times of mourning, clothes made of rough goat-hair, known as sackcloth, were worn (37:34). Because it was so uncomfortable, wearing sackcloth was used to show both grief and, at times, repentance (1 Kings 20:31). Women wore clothing very similar to the men's but with a veil to cover their faces (Genesis 24:65).

For protection, and at times for hunting, the bow and arrow was used (27:3) as were swords and knives. A traveler might use a walking sticks or staff as both an aid for walking as well as a means of defense. Musical instruments such as tambourines and harps were used for entertainment (31:27). Jewelry, such as nose rings, rings, and bracelets were used as ornaments of beauty and as gifts (24:22; Numbers 31:50). Every wealthy man possessed a seal on which was imprinted his own special design to identify what belonged to him (38:18).

The patriarchs ate quite simple food. Bread cakes (14:18; 18:6; 1 Samuel 25:18), soup or stew (25:34), goat's milk and the butter or curds which came from it (18:8; Deuteronomy 32:14; Proverbs 27:27), and on special occasions, meat from either sheep or cattle (18:7) were the normal foods. Water and milk were the most common drinks, and both were stored in either skins (21:19) or jars (24:14). Wine was used to flavor and purify water for ordinary drinking, while less diluted wine was served on special occasions (14:18). Olive oil was also used for cooking, eating, and burning in lamps. Vegetables were also available at various seasons of the year, though these were not the preferred diet (Proverbs 15:17; Daniel 1:12). Fruit, especially figs, grapes, dates, and pomegranates, were also eaten when available (Numbers 20:5). When Moses later described the land of Israel and its agricultural resources, he said: "For the LORD your God is bringing you into a good land … a land with wheat and barley, vines and fig trees, pomegranates, olive oil and honey; a land where bread will not be scarce and you will lack nothing" (Deuteronomy 8:7-9).

LESSON 9:
EGYPT

From the time of the patriarchs until the end of Israel's Divided Kingdom (2000-600 B.C.), the people of Israel were almost always in contact with their neighbor to the southwest, Egypt. Our focus in this lesson will be on the land and people of ancient Egypt.

At the end of this lesson you should be able to:
* *Name the three geographical parts of Egypt and its two cultural divisions;*
* *Identify the two most important cities of ancient Egypt and the land to its south;*
* *Relate one event which took place during each of Egypt's four major periods;*
* *Give the name for Egyptian writing; and*
* *Describe a mummy and what Egyptians believed about the after-life.*

A. THE LAND OF EGYPT

The ancient land of Egypt was shaped much like a flower with a long stem. The stem of the flower was the Nile River, which flows over 900 miles (1450 kilometers) in Egypt alone. The flower on top of the stem was the delta region of Egypt. The delta is that wide area where the Nile divides and spreads out before entering the Mediterranean. Just below the flower portion, on the left side of the stem, is a small leaf. This area is called the Fayum (also spelled Faiyum or Fayyum). It was full of lakes and marshes and was therefore as fertile as the land along the Nile River and in the delta. These then are the three main geographical areas of Egypt – the Nile, the delta, and the Fayum.

Egypt was also divided culturally into two main parts – Upper Egypt and Lower Egypt. Upper Egypt is the area to the south or "up" the Nile River. Lower Egypt is the area located north of the Fayum where the Nile empties into the Mediterranean Sea (see the map to the right). The Egyptian pharaoh was called "King of Upper and Lower Egypt," making him one of the world's most powerful men.

The land to the south of Egypt (modern day Sudan) is called Cush in the Bible and was called Nubia by the Egyptians (Isaiah 1:11; Daniel 11:43). The border between Upper Egypt and Cush was near the city of Aswan at the First Cataract (rapids) of the Nile. At the height of its power, Egypt extended its reign far into Cush, where it obtained valuable trade items. The Egyptians also used Nubian slaves to assist as warriors and as laborers in their monumental building programs. The two most important cities in Egypt were Thebes in Upper Egypt and Memphis in Lower Egypt. The land of Goshen, where the children of Israel lived during the 400 years of their captivity, was probably somewhere along the eastern part of the delta region (Genesis 47:6).

Map 10: Ancient Egypt

In Lesson 2 we learned that "Water is life" in the Near East. In Egypt, the Nile provided a constant and endless source of good water. Except for mining operations, all other Egyptian activity took place within a few miles of either side of the Nile, the source of their life; the rest was desert. Fed by the rainforests of central Africa, the Nile flooded its valley every year, bringing along fresh new topsoil. One could hardly imagine that the Nile would ever dry up (Isaiah 19:7), or even that it could have a seven-year drought (Genesis 41:25-32). Because of the Nile, Egypt had food even when its neighbors in Israel had famine (Genesis 12:10; 41:57).

B. EGYPTIAN HISTORY

Egypt's civilization dates back over 5000 years. Already in 3000 B.C., Egypt was ruled by a single king who was called the pharaoh. Later Egyptian historians divided the pharaohs of Egypt into 30 different family groups known as dynasties. About 3100 B.C. the First Dynasty united Upper and Lower Egypt under one "Ruler of the Two Lands" for the first time.

The first great period of Egyptian history extended from the Third Dynasty through the Sixth Dynasty (about 2700 B.C. to 2100 B.C.). This period is known as the Old Kingdom. It was the time when the first great pyramids were built. During this period, Egyptians began thinking of the pharaoh as a god and as the absolute ruler of all things; his people were considered his slaves. The pharaohs used slave labor to build giant tombs for themselves. Even today people around the world know about these great pyramids. The largest, the Great Pyramid of Khufu (or Cheops), was built about 2600 B.C. It was 480 feet (145 meters) high and took 20 years and over 100,000 men to complete. Some of its thousands of stones weighed hundreds of tons and had to be moved from distant quarries before being tightly assembled. Thus these early Egyptians became some of the greatest builders of all time. The purpose of the pyramid was to provide a secure burial site for Egypt's supreme ruler and god, the pharaoh. Elaborate chambers were laid out within each pyramid to provide the pharaoh with all he would need for the afterlife and also to safeguard him against anyone who tried to rob his tomb. Originally, Egyptians believed that only the pharaoh and his family were worthy of entering the afterlife. Over time the other Egyptians also began building tombs to preserve their own bodies for the next world. The size of the tomb reflected the wealth and social position of the person buried there.

After the Sixth Dynasty, Egypt was weakened by a period of inner turmoil which lasted about 100 years. During this century the Seventh through the Tenth Dynasties fought for control. A famine brought Abraham to Egypt during this unsettled period (Genesis 12:10). With the appearance of the Eleventh Dynasty, a new period of centralized power began, which we call the Middle Kingdom (2125-1675 B.C., Dynasties 11-13). Egypt began spreading its influence into both Cush and Palestine. During the Middle Kingdom, Joseph became first an Egyptian slave and then an assistant to the pharaoh. He used his influence to save his family from the famine which later engulfed the Near East (Genesis 39-47). Egypt also became more involved in international trade during the Middle Kingdom period; through the port of Byblos in Phoenicia, Egypt carried on trade with far away Mesopotamia and with the Minoans on the island of Crete.

After the Thirteenth Dynasty, there was another breakdown in centralized power. Asiatic peoples known as the Hyksos invaded Egypt and established control for more than a century. With the coming of the Eighteenth Dynasty in about 1575 B.C., peace and order were restored and a new era began, called the New Kingdom. Under the rule of the Eighteenth to Twentieth Dynasties (1575-1075 B.C.), Egypt experienced its greatest power and glory, extending its control over Cush, Palestine, and even parts of Mesopotamia. In the midst of their expansion, Moses led the children of Israel out of slavery and into the Promised Land (c. 1450 B.C.). Even after the destruction of pharaoh and his army at the Red Sea, Egypt was once again able to resume its power and to continue as a mighty nation.

Dates (B.C.)	**Periods**	**Dynasties**
3100 - 2200	Old Kingdom	1 - 6
2200 - 2050	1st Intermediate Period	7 - 10
2125 - 1675	Middle Kingdom	11 - 13
1675 - 1575	2nd Intermediate Period	14 - 17
1575 - 1075	New Kingdom	18 - 20
1075 - 700	3rd Intermediate Period	21 - 25
700 -300	Late Period	25 - 31

Chart 3. Main Periods of Egyptian History

After yet another period of weakness (1075-700 B.C.) came the final period of ancient Egyptian independence, known as the Late Period. It lasted from 700-300 B.C. Over this time Egypt gradually lost its power and became a "broken reed" able to be invaded by the Assyrians (Isaiah 36:6). This loss of power and prestige was the result of God's judgment: "I will set fire to Egypt...Thebes will be taken by storm, Memphis will be in constant distress" (Ezekiel 30:16). Godlessness resulted in Egypt's permanent decline.

C. EGYPTIAN WRITING

About the time that the Sumerians were developing cuneiform in Mesopotamia, the Egyptians were also experimenting with ways to write their language. Their first successful attempt was a picture language that we call hieroglyphics. Originally, a hieroglyphic picture stood for an entire word, but through further developments, the pictures came to represent syllables and individual sounds as well. Thus, a very sophisticated system of writing arose. However, it was also very complicated, so only a small percentage of Egyptians ever learned to read and write it. It even took modern scholars until well into the nineteenth century before they finally learned to read this ancient language. Now that researchers can read the writing on ancient Egyptian documents, stones and buildings, we can understand much more about Egyptian history, life, and culture.

D. LIFE AND SOCIETY IN ANCIENT EGYPT

The pharaoh and his family were at the top of Egyptian society. Besides his duties as king, the pharaoh served as the highest judge of the land and as high priest at the major religious festivals and sacrifices. He was even adored as a god. Next to the pharaoh were the priests. They were trained not only in religion but also in numerous other types of learning. Because of their educational background, the priests also became the doctors, lawyers, scientists, professors, and general problem solvers of their time (Exodus 7:11).

Just beneath the priests were the nobles. Nobles served as army officers, diplomats, and royal advisors and held other important posts in government. Like the priests, the nobles were usually quite wealthy. Potiphar, the man who bought Joseph and appointed him to oversee all his property, was one of those nobles (Genesis 39:1). Over the past century archaeologists have discovered some of their beautiful and elaborate tombs. Craftsmen also played an important role in Egyptian society. They worked in wood, metal, and stone and were in charge of a variety of building and artistic work. Not as wealthy as the other upper classes, they nonetheless were able to live in cities and enjoy some of the amenities of urban life.

At the bottom, making up the vast majority, were the peasants and common slaves. The peasants were the farmers and laborers of Egypt. Without them the other classes would have starved, yet they remained poor and were treated poorly. Economically, they were barely more than slaves. They worked hard in their fields with only a few animals and primitive tools to help them. In times of war, they were forced into the army, and during the periods of great building projects they were the ones who provided the sweat and muscle to erect great monuments. Life as a peasant or slave was filled with difficulties and drudgery. However, slaves that had special education or talent, like Joseph, could rise to some importance (Genesis 39:2ff.).

E. EGYPTIAN RELIGION

Egyptian religion was complex and involved the worship of many gods. Ptah was the god who fathered all of the other gods and created the world. Thoth, the god of wisdom, laws and sacred writings, was usually pictured as an ibis bird or baboon. Re, the sun god, was worshiped as lord of earth and sky; he was often pictured as a falcon-headed man. There was also a family – Isis the mother, Osiris the father, and a child Horus. Osiris was killed and then became god of the underworld. Horus, pictured as a falcon, was the sky god and took on human form as the ruling pharaoh. That is why the pharaoh was worshiped as a god and was expected to serve as a mediator between the various gods and men. Elaborate myths were told about these and a multitude of other gods. As in Mesopotamia, each city revered special gods, and each household had its own special gods which they honored. The Egyptians worshiped almost anything, from the sun to the Nile and from animals to imaginary creatures.

In order to keep in favor with all these gods, the Egyptians filled their land with magnificent temples. Hundreds of priests carried out the required rituals and sacrifices. The use of magic and superstition was also mixed with the worship of the gods (Exodus 7:11ff). Religion played a vital role in Egypt.

The Egyptians also had a strong belief in a life after death. They were almost obsessed with preparing for death and the afterlife. They believed in a judgment and bodily resurrection. Osiris, the god of death, weighed each heart to judge its worthiness to enter eternity. The kind of life you lived here, together with your status in this life, would tip the scales for or against you in the life to come. If a person were good and honored in this world, he would gain eternal happiness and respect. If not, he could expect punishment and shame. To preserve the human body for this afterlife, the Egyptians made the bodies into mummies. Special chemicals were applied to dry out the bodies, and the inner organs were removed and placed in special jars. The bodies were then wrapped in papyrus, put into airtight coffins, and buried after a period of mourning, which lasted up to 70 days for important members of society. When Jacob died in Egypt, his body was mummified as was the custom, and the Egyptians mourned him for 70 days (Genesis 50:2-3).

Egyptians believed that after death a person's soul was released from the body, yet at night, it would return to rest with its mummified body or a statue of the person. Wealthy people built elaborate tombs with exquisite decorations. Food, along with models of their houses, farms, animals, tools, luxuries, and even slaves were then placed in the tomb. They thought that all these things would accompany the person to the next life. Instructions for getting to the next world were engraved on the walls of the tomb and written on papyrus scrolls called *The Book of the Dead* which was then placed next to the mummy. They felt that doing all this would help the dead person arrive alive and well in the next world. Written curses threatened anyone who might disturb the grave. Unfortunately, little deterred the thieves, and eventually most tombs were broken into and the precious objects stolen. A few tombs did remain untouched until the 20th century. One of these was the burial place of the boy king Tutankhamen, often called King Tut. The treasures found in his tomb, now on display in Egypt, make it very clear how seriously the Egyptians took their preparations for eternity.

LESSON 10:
THE EXODUS AND THE BUILDING OF THE TENT OF MEETING

After they had spent 400 years in Egypt, the LORD in his mercy sent Moses to lead the people of Israel back to the Promised Land. In Lesson 10, we will learn about when this took place and the route which God's people probably followed as they journeyed from Egypt to Palestine. We will also study the Tent of Meeting which God instructed Israel to build as his dwelling place among his people.

At the end of this lesson you should be able to:

* *Give the approximate date for the Exodus from Egypt;*
* *Name and briefly describe the three types of laws God gave to Moses on Mt. Sinai;*
* *Give a description of the Tent of Meeting and its courtyard, noting what was in each;*
* *Identify the clothing worn by ordinary priests and the high priest;*
* *List the three duties of the priests; and*
* *Explain the significance of the Tent of Meeting for the community of Israel.*

A. THE ROUTE OF THE EXODUS

Although the Bible mentions the names of many places along the route of the Exodus, we are unable to determine exactly where most of those places were. These place names have changed over the centuries and are now almost impossible to pinpoint. We do know that the LORD led the Israelites from the land of Goshen to the Red Sea (or Sea of Reeds; cf. footnote on Exodus 13:18 NIV). After crossing the Red Sea, they then traveled south along the western edge of the wilderness of Shur (Exodus 15:22), and on through the Desert of Sin (Exodus 16:1). There God began to make a special form of food fall from heaven in the morning. It was called "manna." From the Desert of Zin, the Israelites continued into the Desert of Sinai (Exodus 19:1) and camped at Mt. Sinai. After a stay of almost two years during which they received God's Law and built the Tent of Meeting, they moved north to the Desert of Zin, stopping at Kadesh-Barnea (Numbers 20:1). There the Israelites sent out spies and rebelled against God's command to enter the land. Because they did not believe in God's protection, God required them to wander for 40 years in the wilderness before they were allowed to enter the Promised Land (Numbers 14:20-25; Deuteronomy 29:5). Map 11 shows the probable location of these areas.

B. THE DATE OF THE EXODUS

1 Kings 6:1 tells us, "In the four hundred and eightieth year after the Israelites came out of Egypt, in the fourth year of Solomon's reign over Israel…he began to build the temple of the LORD." By using other biblical dates and explanations, we can accurately date the fourth year of Solomon's reign to about 966 B.C. This means that the Exodus took place in 1446 B.C. For our purposes, the approximate date of 1450 B.C. will be sufficient.

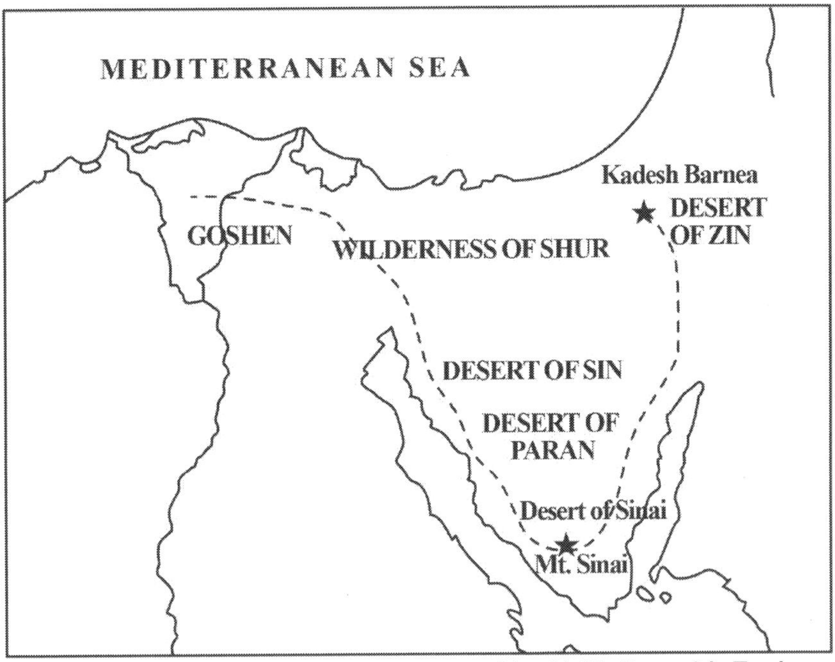

Map 11. The Route of the Exodus

For a number of reasons, many modern scholars object to this date for the Exodus. Most claim that the Exodus must have taken place almost two centuries later, sometime in the 1200s B.C. They have developed an elaborate system of arguments to support this theory. But because Scripture clearly points to a date of about 1450 B.C., we accept this as the actual date. Although Egyptian scholars argue about the dating of the various pharaohs, it seems that Amenhotep (or Amenophis) II was the pharaoh in 1446 B.C.

C. GOD GIVES HIS LAW TO MOSES AND THE PEOPLE OF ISRAEL

The first great act of God following his miraculous deliverance of Israel from Egypt was the giving of his Law on Mt. Sinai. We do not have the space in this book to study the Mosaic Law in detail; we will only mention the three main types of laws given at Sinai—the moral laws, the ceremonial laws, and the civil laws.

God's moral laws are summed up in the Ten Commandments and express God's will for all people of all times. Their purpose is to show us our sinfulness in thought, word, and deed, and thus our desperate need for a Savior from sin and its consequences.

The ceremonial laws given at Mt. Sinai instructed Old Testament believers how to worship God properly. They included rules regarding proper observance of the Sabbath, commands for bringing sacrifices, instructions about religious festivals and their observance, and finally guidelines for the priests and their priestly duties. The ceremonial laws were intended only for God's Old Testament people and were abolished by Christ with his fulfillment of the Law. Since the Old Testament ceremonial laws were only temporary commands and were not repeated in the New Testament, God tells us that we are now free to worship him with whatever forms and practices are most beneficial and suited for Christians today (Colossians 2:16-17; John 4:21-24).

The civil laws were also only intended for God's Old Testament people and consisted of rules and regulations concerning daily life for the people of Israel. As the name implies, the civil laws governed Israel as a nation and dealt with crimes such as injury to another person or damage to property. God gave these laws to Israel to help their society run smoothly while they were a nation. With Christ's coming and the dispersion of the Jewish nation following Christ's death, these civil laws came to an end.

D. THE BUILDING OF THE TENT OF MEETING

"I am the LORD and I will redeem you with an outstretched arm and with mighty acts of judgment. I will take you as my own people, and I will be your God" (Exodus 6:6-7). This was God's covenant with his people. God did make Israel his special people, and gave them commands on how to live and act in a way pleasing to him. To show his people that he himself was also fulfilling the rest of that covenant, "that he would be their God," the LORD gave them instructions for building a place where he could make his presence among them known – the Tent of Meeting. God told Moses: "Have them make a sanctuary for me, and I will dwell among them" (Exodus 25:8). When Israel had finished building the Tent of Meeting, God descended in the form of a cloud and dwelt in the midst of his people (Exodus 40:34). The Tent of Meeting served, therefore, as a physical reminder that the LORD lived among his people and was always there to help and to save them in their times of need. Centuries later God would fully live among men in the person of Jesus Christ (John 1:14).

The Old Testament book of Exodus describes in great detail how the Tent of Meeting was to be constructed. It consisted of two main parts: the courtyard and the inner tent. The courtyard was a large area, 150 feet (46 meters) long and 75 feet (23 meters) wide, surrounded by a fence of linen curtains 7.5 feet (2.3 meters) high. This fence set apart God and his holy place from the rest of the camp of Israel. When the Israelites came to worship or bring sacrifices, they would enter through a gate of the courtyard on its eastern end. Only circumcised Jews were allowed to enter this area.

Inside the courtyard was an altar for burnt offerings, a bronze basin, and the inner tent, also referred to as the Tent of Meeting. The altar of burnt offerings was a square platform 7.5 feet (2.3 meters) long on each side and 4.5 feet (1.4 meters) high. It was made of acacia wood and covered with bronze. On each of its four corners there was a horn. Upon this altar, the priests made sacrifices every morning and evening as well as for special occasions. Beyond the altar, near the Tent of Meeting, stood a great bronze basin in which the priests ceremonially washed before conducting their duties.

The Tent of Meeting (also called the Tabernacle in some English translations of the Bible) was the center of the entire Tent of Meeting structure. It was 45 feet (14 meters) long, 15 feet (4.6 meters) wide, and 15 feet (4.6 meters) high. The Tent of Meeting was made of precious woods and skins, and it was divided into two parts by a curtain. The larger of its two areas was designated as the Holy Place, and the smaller room in back served as the Most Holy Place. The entrance to the tent was also on the east side. Only priests were allowed to enter the tent. Walking through the curtain, a priest would find himself in the Holy Place. To the right there was placed a table for the bread of the Presence (or *showbread*). The priests placed twelve fresh loaves of unleavened bread there each Sabbath and then ate the old ones. These loaves represented the twelve tribes of Israel. To the left was a golden lampstand with a lamp on each of its seven branches. It was made of solid gold. Its light symbolized God's presence. Straight ahead, in front of the curtain to the Most Holy Place, was the incense altar. It was a rectangle of about 1.5 feet (0.5 meters) in length and width, was about three feet (1 meter) tall, and was made of acacia wood and covered in gold with a horn at each corner. Every morning and evening, the priest tended to the lamps and then used fire from the altar of burnt offerings to offer up incense on the incense altar. The burning incense symbolized the prayers being offered up by the priest to God (Revelation 5:8).

Only the high priest could pass through the curtain and enter into the Most Holy Place, and he could do that only once a year—on the Day of Atonement (Leviticus 16). Inside this room was the precious Ark of the Testimony (or Covenant). It also was made of acacia wood and covered in gold, and it supported the atonement cover which was topped with two golden cherubim (angels) spreading their wings. Inside the Ark of the Testimony were kept the two tablets of the Law given to Moses, a jar full of manna, and the staff of Aaron which had sprouted and blossomed (Hebrews 9:3-4). There in the Most Holy Place resided the glory of the LORD as he lived among his people (Exodus 25:8, 22).

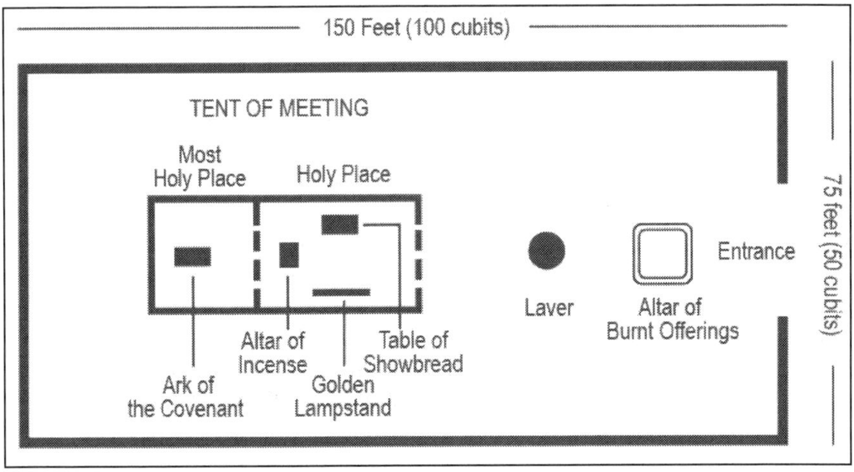

Chart 4. The Tent of Meeting

E. THE PRIESTS

God appointed Aaron and his descendants to serve as priests (Exodus 28:1ff). God gave them the exclusive rights to conduct the worship services for God's people. They also wore special clothing to indicate their special role. Their priestly garments included a long white robe with a colorful belt, linen pants, and a special hat. In addition to this, the high priest wore a long blue overcoat with bells and cloth pomegranates on its bottom fringe. Over this robe he wore an ephod – two pieces of linen joined at the shoulder to form a vest. The high priest also wore a breastplate containing 12 precious stones with the names of each of the tribes of Israel inscribed on it. On his tall hat was written "Holiness to the LORD."

Besides officiating at the Tent of Meeting sacrifices (Leviticus 1-7; 16-20), the priests were in charge of teaching the people God's Word (Leviticus 10:8-11). They were also commanded to set good examples for others by the way in which they lived (Leviticus 21). Finally, the high priest was also in charge of the Urim and Thummim. Scholars disagree on what form the Urim and Thummim took, but it is clear that the high priest used them to consult with God and to assist in making God-pleasing decisions on behalf of God's people (Exodus 28:29-30).

LESSON 11:
CONQUESTS ON THE EAST BANK

In Lesson 10, we traced the route of the Exodus up to the Israelites' arrival at Kadesh-Barnea. They still wandered the desert for the next 38 years as punishment for their disobedience and rebellion. In this lesson, we will learn what happened at the end of their years of wandering. We will trace their journey through much of the East Bank and see how they avoided some peoples but fought and defeated others.

At the end of this lesson you should be able to:

* *Name the three peoples that the Israelites did not fight against and explain why they did not;*
* *Identify the two kings and peoples that Israel conquered;*
* *Locate on a map the areas where these five peoples lived; and*
* *Trace the approximate route the Israelites took as they passed through the East Bank.*

A. THE WAY TO THE PROMISED LAND

When Moses led God's people out of Egypt, the Lord told him the route they were to travel. But God did not have Moses lead his people directly to the Promised Land along the main road by the Mediterranean Sea. This would have led them past some of Egypt's strongest army camps as well as the country of the warlike Philistines. These trials would have been too difficult for the Israelite nation at this stage in their development. As God himself noted, Israel would have been too afraid and would have turned back in the face of such opposition (Exodus 13:17). By making them take a longer and less threatening route (shown on Map 12), God was able to strengthen his people's trust in him and thus better equip them to confront their enemies.

Eventually they did reach the border of the Promised Land near Kadesh Barnea, but their faith had not yet matured. They did not follow the advice of the two good spies to go forward into the land. Their excuse was that they were concerned for the welfare of their children (Numbers 14:3). Because of their lack of faith, only their children (and the two faithful spies,

Joshua and Caleb) survived to enter the land of Israel. Forty years after leaving Egypt, Israel returned to the borders of the Promised Land and entered from its eastern borders.

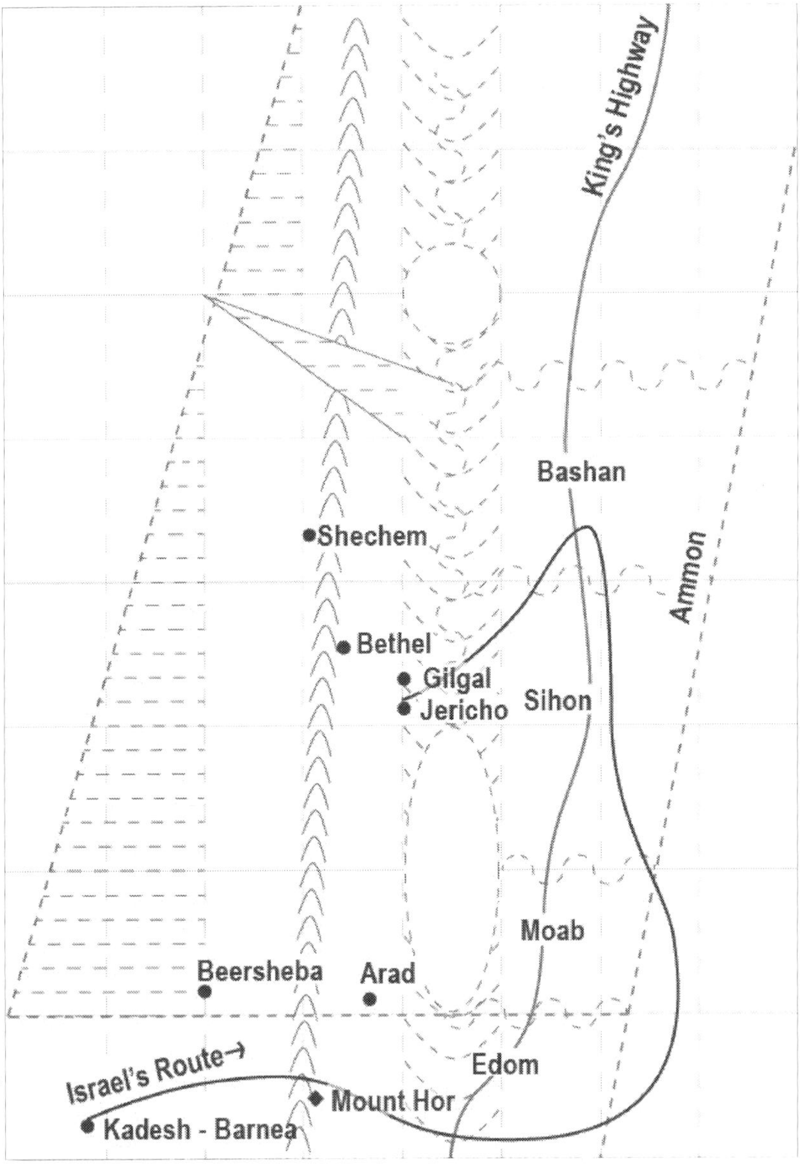

Map 12. Israel's Travels and Conquests on the East Bank

Once they left Kadesh Barnea, Israel traveled east to Mt. Hor (its exact location is today unknown). There Aaron died and the king of the Canaanite city of Arad attacked Israel, but Israel quickly defeated him (Numbers 20:22; 21:1ff.). In spite of their victory, the impatient Israelites soon forgot the Lord's blessings and began to grumble. God responded by sending a plague of poisonous snakes. God graciously saved many Israelites, however, through the miracle of the bronze snake (Numbers 21:4-9). After this incident, they once more continued their journey towards Edom.

B. EDOM

During the period of the Exodus, the people of Edom lived in the area south of the Zered River between the Rift Valley and the desert's edge. The rugged mountain range that filled this area was called Mt. Seir, and this name was sometimes used to refer to the whole land of Edom (Deuteronomy 1:2).

The Edomites were descended from Esau so they were relatives of the Israelites (Numbers 20:14). Genesis 36 mentions many important rulers and cities of the Edomites. Outside of what the Bible records about these descendants of Esau, we know very little. We do know that as Moses led Israel closer to their final goal, he wanted to take Israel through Edom along the main eastern road, called the King's Highway (Numbers 20:17). This main trade route ran through the East Bank all the way from Mesopotamia to Egypt.

This route not only was the easiest way to cross this area, but it also had good supplies of water and food. Thus Israel requested permission from the Edomites to travel through their land. They promised that they would travel only on the main highway, pay for any food or water, and not cause any trouble (Numbers 20:14-19). But the king of Edom refused their request and threatened to attack them if they entered his territory (Numbers 20:18-20).

Because of this Israel did not go through Edom along the King's Highway. Instead they bypassed Edom. It is unclear whether they went around the eastern or western side of Edom, but they probably traveled along the edge of the desert. When they came to the Zered River valley, they camped (Numbers 21:12). They were then at the border of Moab.

C. MOAB

Genesis 19:37 tells us that the Moabites were also closely related to Israel. They were the descendants of Lot's oldest daughter. Their language was similar to the Hebrew language of the Israelites, and, like Israel, they were well known as sheep herders. They also were involved in trade along the King's Highway which ran through the middle of their country. At some periods of history Moab controlled the entire eastern shore of the Dead Sea. During the period of the Exodus, however, the Amorites under King Sihon had pushed the Moabites south of the Arnon River. Despite their ties to Abraham's family, the Moabites were not worshipers of the true god. Instead they worshiped their "detestable god," Chemosh (1 Kings 11:7) to whom they offered human sacrifices (2 Kings 3:27)!

God told the Israelites not to harm the Moabites because they too were Israel's relatives. Just as with Edom, Israel was compelled to travel around the land of Moab. The Lord commanded them not to take any land from either of these cousin nations (Deuteronomy 2:8-9). So Israel marched around the land of Moab and set up camp in the Arnon River Valley, the border between the Moabites and the Amorites (Numbers 21:13-14).

Two additional references to the Moabites should be noted. Fearing Israel's power, the Moabites called upon the sorcerer Balaam (who is also mentioned in other ancient writings besides the Bible) to come and curse the Israelites (Numbers 22-24), but God then allowed Balaam only to bless his people. Later, the Moabites, together with the Midianites (a tribe living among the Moabites and in the area to the east), were able to lure Israel into idolatry through sexual sins (Numbers 25).

D. AMMON

The Ammonites lived along the desert's edge just to the north of Moab. They were the descendants of Lot's younger daughter. Once again God forbade Israel from attacking these relatives (Deuteronomy 2:19). So instead, Moses led Israel west of Ammon through the territory controlled by King Sihon of the Amorites.

Because of their relationship to Israel, God told the Israelites not to conquer the land or people of Edom, Moab, or Ammon.

E. THE AMORITES

Sometimes the Bible uses the word "Amorite," like "Canaanite," to speak of any of the heathen peoples living in the Promised Land (Genesis 3:8). However, the name Amorites was also used to describe a large group of loosely connected tribes settled along the Eastern borders of Palestine. It seems that the Amorites migrated south from Syria into Canaan sometime around 1500 B.C. One group of Amorites under King Sihon crossed the Jordan and took over Moab's territory north of the Arnon River. Israel requested permission to pass through this Amorite territory along the King's Highway, but Sihon refused. Instead, he gathered up his army and attacked Israel. The result was an overwhelming victory for Israel. They took over the land of the Amorites and continued their journey along the road towards Bashan (Numbers 21:21-35).

F. BASHAN

In this period, Og, the king of Bashan, controlled the East Bank territory north of the Jabbok River. This land was famous for its fertile pastures (Micah 7:14), its large and strong bulls (Psalm 22:12), and its oak forests (Isaiah 2:13). At this time the kingdom of Bashan consisted of 60 large and well-fortified cities (Deuteronomy 3:4-5). Their king, Og, was an enormous man who slept in a bed 13 feet (4 meters) long and 6 feet (2 meters) wide (Deuteronomy 3:11). With the Lord's help Israel was nonetheless able to completely overcome this powerful king and his country (Deuteronomy 4:4-11).

These two powerful enemies, the Amorites and Bashan, were defeated by Israel, and their land became Israelite territory.

G. ON THE PLAINS OF MOAB

Just above the Dead Sea the Rift Valley spreads out into a flat wide plain (the Hebrew *kikkar* meaning "circle"). This circular plain spreads out on both sides of the Jordan River at this point. The area on the West Bank was called the plains of Jericho (Joshua 4:13); the area on the East Bank was called the plains of Moab (Numbers 22:1). Israel camped on the plains of Moab before they entered the Promised Land. While camped there, the incidents with Balaam and the immorality with the Moabites and Midianites took place. It was also here that the Lord had Moses take a census of all the people. Afterwards, final preparations were made for that long-awaited day of crossing the Jordan and entering the Promised Land.

LESSON 12: CONQUESTS ON THE WEST BANK

In Lesson 11, we traced the children of Israel as they conquered the lands of Bashan and the Amorites on the East Bank. In this lesson we will describe their journey as they crossed the Jordan River and conquered the West Bank of the Promised Land.

At the end of this lesson you should be able to:
* *Write the Old Testament meanings of the Hebrew word Arabah;*
* *Describe the main battle of Joshua's central and southern campaign and the main battle in his northern campaign, and where each took place;*
* *Name Israel's headquarters on the West Bank; and*
* *Name the three areas of the Promised Land that remained unconquered.*

A. IN THE PLAINS OF MOAB

After conquering the kingdoms of Bashan and the Amorites on the East Bank, Israel camped on the plains of Moab across from Jericho. According to Deuteronomy 1:1 this area was in the *Arabah*. The Hebrew word *Arabah* is used in the earlier part of the Old Testament to describe the whole Rift Valley. In the later books, the word *Arabah* often refers to the part of the Rift Valley south of the Dead Sea (Amos 6:14).

On the plains of Moab, Moses spoke to the Israelites all the words that we have recorded in the book of Deuteronomy. He reminded the people of God's protection during their long journey from Mt. Sinai to the Jordan (chapters 1-3) and how God gave them the Laws of God on Mt. Sinai (4-26). Moses then gave the people instructions on what to do after crossing the Jordan (27) and told them what would happen if they obeyed or disobeyed God's commands (28-30). Finally, he appointed Joshua as his successor (31). Then Moses, after singing a song of praise and prophecy (32-33), went to the top of Mt. Nebo where he died and was buried by God (34). This ends the Pentateuch, the first five books of the Old Testament, which Moses wrote. Deuteronomy 34, which describes Moses's death, was written later as an appendix to the book.

B. CROSSING THE JORDAN

Joshua then led the Israelites across the Jordan and set up camp at Gilgal (Joshua 3:1-5:9). Gilgal became the headquarters for the Israelites during the coming campaign. The women and children were probably left there to take care of the herds while the men went off to fight against the Canaanites. The main battles which followed can be divided into two parts. First, there was a campaign in the central and southern sections of Palestine. Then, after conquering these areas, Joshua also led the army on a northern campaign. Through these two campaigns, Israel gained control over much of the West Bank. Follow along on Map 13 as you read the next sections.

C. THE CAMPAIGN IN CENTRAL AND SOUTHERN PALESTINE

Joshua 6-10 describes Israel's southern and central campaign against the Canaanites. It began with the miraculous overthrow of Jericho (Joshua 6). Because one of the Israelites, Achan, disobeyed God's orders and kept some of the plunder, Israel was defeated in its first attack on the city of Ai. However, after Achan was discovered and punished, God granted them victory over this city as well (7-8). The people of Gibeon then deceived Israel; as a result they were spared from destruction but became slaves to the Israelites instead (9).

The other Canaanite kings thought that Gibeon should be punished for making peace with Israel, so they formed an alliance and went to attack their city. The people of Gibeon appealed to Joshua for help, and the army of Israel came to Gibeon's defense. The battle took place in the Valley of Aijalon outside Gibeon. There God gave the Israelites a total victory, even making the sun stand still so the Israelites could finish off their enemies. The Canaanite kings who fought in this battle all fled to Makkedah and hid in a cave where they were discovered and killed. This was the decisive battle of the campaign (Joshua 10).

With the fiercest fighting ended, Joshua and his men moved on and destroyed the last of the opposition in each of the towns of the defeated kings. Working counter-clockwise, Joshua led the army southwest through the highlands and foothills and conquered the cities of Libnah, Lachish, and Eglon, one after another. Then he turned east and headed back up the watershed, conquering Hebron and Debir before returning to Israel's headquarters at Gilgal. This brought to a conclusion the central and southern campaign.

D. THE NORTHERN CAMPAIGN

The Canaanite tribes living in northern Palestine became afraid when they heard that the Israelites had totally defeated their southern neighbors. The King of Hazor sent messages to all the other northern kings urging them to form a coalition against the Israelites. The kings along the seacoast, in the highlands and mountains, and those near the Sea of Galilee all rose up against the army of Israel. They fought the crucial battle at the Waters of Merom (Joshua 11:7). The Israelites were once again victorious. Following God's directive, the army pursued its enemies in all directions "until no survivors were left" (Joshua 11:8). Hazor, however, was the only city that Joshua destroyed. The Israelites occupied the rest of the cities of northern Palestine, killing only those inhabitants who resisted.

E. ISRAEL TAKES CONTROL OF THE WEST BANK

Israel now controlled the land God had promised to Abraham. In Joshua 11:16-17, the writer describes the large area of their conquests:

> So Joshua took this entire land: the hill country, all the Negev, the whole region of Goshen, the western foothills, the Arabah and the mountains of Israel with their foothills, from Mount Halak which rises towards Seir, to Baal Gad in the Valley of Lebanon below Mount Hermon.

In Joshua 11:18, however, we also read that Joshua "waged war against all these kings for a long time." The two campaigns described above broke the power of the Canaanites. However, it still took many years for Joshua and his army to overcome the remaining survivors and to take command of all the smaller cities and tribes scattered about. It is also clear from the biblical description that not all of Canaan was under Israel's control. From Joshua 13 we see that there were three main areas which remained unconquered:

1. The plain of the Philistines (Joshua 13:1-3);
2. The northern part of the Plain of Sharon and the Acco plain (13:4-5);
3. The East Bank above the Sea of Galilee (13:5).

During the period of the judges, tribes and peoples from these areas continued to cause problems for the Israelites. Only in the time of King David were all of these areas finally conquered.

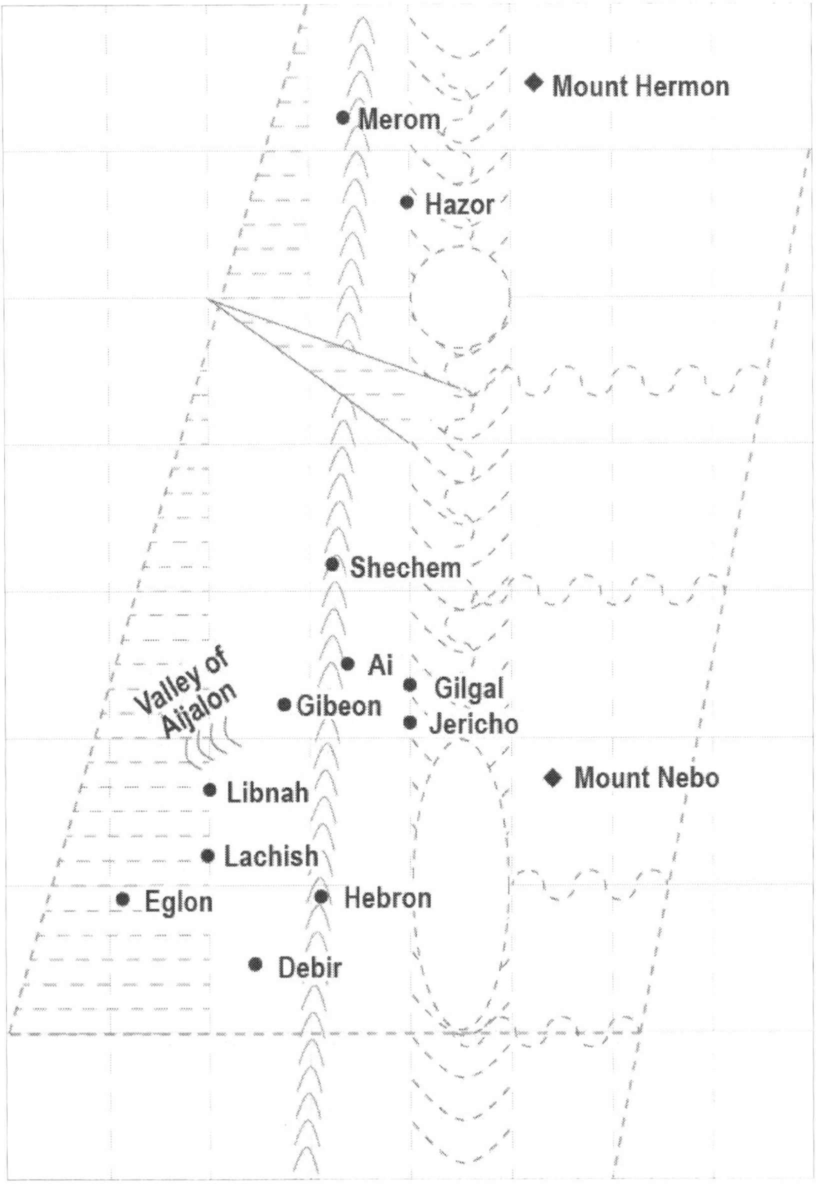

Map 13. Israel's Conquests of the West Bank

LESSON 13:
THE TWELVE TRIBES AND THEIR LAND

The children of Israel had taken possession of the Promised Land. Under God's guidance and direction, the land was divided among them, first to each of the 12 tribes, and then, to every family within each tribe. That division of the land will be the focus of this lesson.

At the end of this lesson you should be able to:
* *Locate which area of the Promised Land each of the 12 tribes received;*
* *Tell which of the 12 sons received a double portion of land and which received none;*
* *Tell which tribe received land on both the East and West Banks; and*
* *Identify which tribe was later mixed with another, and which one moved.*

A. THE TWELVE TRIBES

Israel was the special name God gave to Jacob after Jacob had wrestled with him along the banks of the Jabbok River at Peniel (Genesis 32:28). Jacob's 12 sons were destined to become the fathers of the original 12 tribes of Israel. The sons of Jacob who journeyed down to Egypt under the protecting hand of their brother Joseph are listed in Genesis 46:8-25. They are also listed in the left column in the chart below.

By the time the children of Israel had arrived back in the Promised Land, the make-up of the 12 original tribes had changed. Jacob had prophesied about this change when he told Joseph about God's plans for Joseph's two sons Manasseh and Ephraim:

> Your two sons, born to you in Egypt before I came to you here, will be counted as mine. Ephraim and Manasseh will be mine just as Reuben and Simeon are mine. But any children born to you after them will only be counted as yours, and in the territory they inherit they will be counted under the names of their brothers. (Genesis 48:5-6)

So instead of Joseph's name being recorded as one of the 12 tribes, his two sons – Ephraim and Manasseh – were each counted as a tribe in Israel.

So why does the Bible continue to speak of 12 tribes instead of 13? The Lord took the tribe of Levi and put them in charge of caring for his Tent of Meeting (Numbers 4-5). As a result, they were no longer counted among the 12 (Numbers 26:62), and they did not receive a portion of the land of Israel like the other tribes. Instead the Lord gave them 48 cities scattered throughout Israel (Joshua 21). The priests of the tribe of Levi lived with their families in those cities while giving special service to the Lord.

Jacob's 12 Sons (Genesis 46)	12 Tribes Receiving Land (Joshua 13-21)	Area Each Received
1. Reuben	Reuben	East Bank from the top of the Dead Sea to the Amon
2. Simeon	Simeon	Negev west of the watershed
3. Levi	**Manasseh**	Gilead (northern Bashan) **and** northern Samaria
4. Judah	Judah	Judea (West Bank west of the Dead Sea)
5. Issachar	Issachar	Jezreel Valley east of the watershed
6. Zebulun	Zebulun	Jezreel Valley east of the watershed
7. Gad	Gad	East Bank on both sides of the Jabbok
8. Asher	Asher	Plain of Acco and Highlands of Galilee west of the watershed
9. Joseph	**Ephraim**	Southern Samaria just north of the Corridor
10. Benjamin	Benjamin	The eastern Corridor
11. Dan	Dan	The western Corridor
12. Naphtali	Naphtali	Highlands of Galilee east of the watershed

Chart 5. The Twelve Tribes of Israel

Thus, when Israel conquered the Promised Land, 12 tribes received areas of territory, and the tribe of Levi settled in their 48 cities. Joshua 13-21 recounts how the land was divided up among the 12 tribes. Map 14 shows the approximate division of the land. You will notice that, in general, the borders between each of the tribes follow the natural divisions and geographical areas that we have already studied. These natural borderlines allowed each tribe to identify its own borders and those of its neighbors, and this helped minimize border disputes. Although the land for most tribes was decided by lot, God first directly assigned areas in the heartland to the tribes of Joseph's sons (Ephraim and Manasseh) and to Judah. This emphasized the central role of these tribes in God's plan.

B. CHANGES TO THE DIVISION

Note that there are still 13 portions on our map, however. Manasseh, Joseph's firstborn, received a double portion of land – a part on the East Bank and a part on the West Bank. Thus, although the land was divided into 13 different areas, these areas were still only divided among 12 tribes. It should also be noted that some of the tribes did not fully occupy the portions of land that each received.

In our previous lesson, we mentioned that there were three areas which remained unconquered until the time of David. One of those areas was the land of the Philistines along the southwest coast of Israel. The presence of the Philistines in this part of the land kept the tribes of Simeon, Judah and Dan from occupying all of their territory. In fact, as time passed, the tribe of Simeon was almost completely absorbed into the tribe of Judah. This fulfilled Jacob's prophecy (Genesis 49:6-7) that both Levi and Simeon would be scattered throughout Israel as a result of their sin (see Genesis 34). Also, the tribe of Dan eventually relocated to the most northern area of the Rift Valley, just north of the marshy area called the Huleh Basin (the area of the Rift Valley just north of the Sea of Galilee).

Because of the presence of the Phoenicians, the other tribes which lived further north along the coast of the Mediterranean also had difficulty in completely occupying their land. In addition, Manasseh had trouble claiming all of its land along the East Bank. Though fighting continued with numerous groups of Canaanites, the Israelites firmly controlled the majority of the land. Thus God had fulfilled his promise to Israel's forefathers – Abraham, Isaac, and Jacob. He took their descendants out of Egypt and gave them Palestine as their homeland. As time passed, the tribes merged with each other within the four major geographical areas of the heartland which shared similar soil and agriculture. These areas were the highlands of Judea, the highlands of Samaria and the highlands of Galilee on the West Bank, and Gilead on the East Bank.

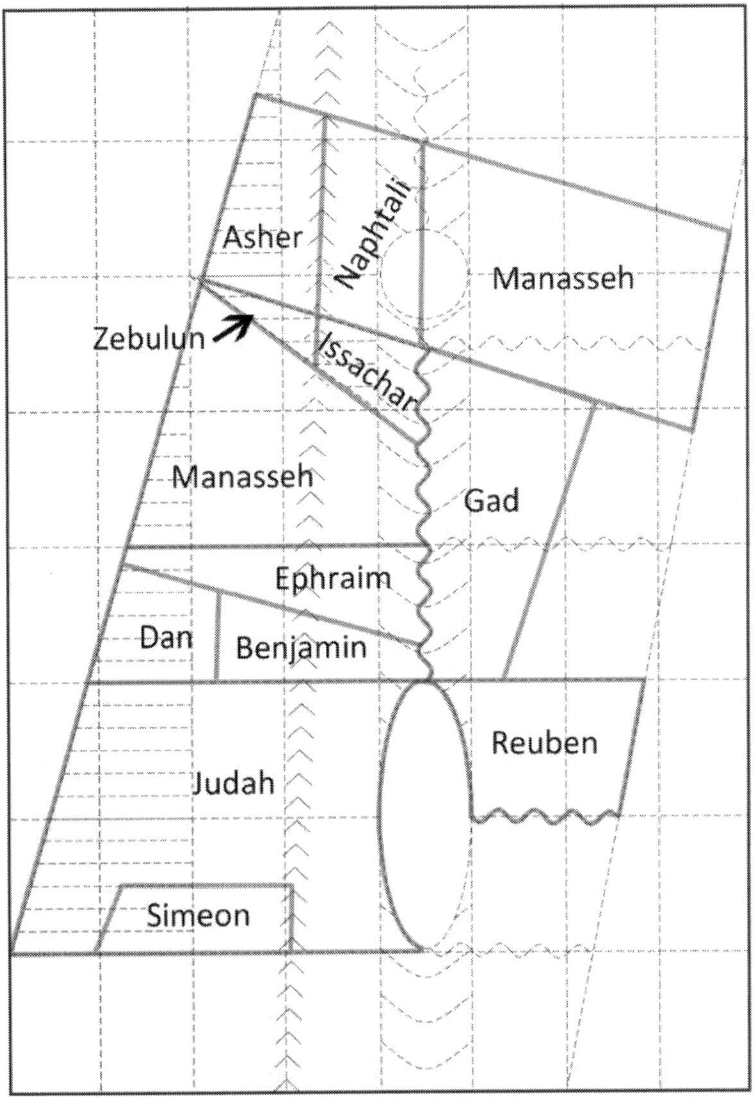

Map 14. The Division of Israel Among the Twelve Tribes

LESSON 14:
THE HEBREW YEAR:
ITS WEATHER, CROPS, AND FESTIVALS

The weather in Israel is similar to that of many subtropical regions. In this lesson, we will take an in-depth look at Israel's weather and growing seasons. We will see how the various seasons of the year fit in with the major religious festivals that God established for his people.

At the end of this lesson you should be able to:

* *Name the four types of moisture that Israel received;*
* *Name Israel's hottest and coldest months;*
* *Tell when grain is planted and harvested, and when olives and grapes are harvested in Israel;*
* *Identify the three important feasts during which every Hebrew man was required to appear before the Lord; and*
* *Explain how those festivals were related to the harvest seasons in Israel.*

A. THE HEBREW YEAR

The Exodus from Egypt began a new period in history for the people of Israel. The establishment of the Passover was to be a constant reminder of that new beginning. To demonstrate this change even further, God commanded the Jews to change their calendars. The month of the Exodus was now to be counted as the first month of the new year (Exodus 12:1-2). This month was called *Abib* or *Nisan*. It began just before the beginning of spring and varied slightly from year to year because the Hebrew calendar was based on the rising of the moon. Each 28- or 29-day cycle of the moon formed one Hebrew month. Thus their calendar year had 12 months of 28 or 29 days (and about every other year they had a thirteenth "leap" month). The names of the Hebrew months, taken from the Babylonians, are found in Chart 6.

B. THE WEATHER IN ISRAEL

Israel has just two main seasons—the dry season and the rainy season. For most of the year the weather in Israel is quite hot. However, the weather can vary considerably from region to region. Along the seacoast the weather is often quite different from that of the highlands. Likewise, the weather in the Rift Valley varies greatly between Galilee and the Dead Sea.

The right column of Chart 6 indicates the general weather patterns in the land of Israel. During the months of September and October, the weather begins to change from hot to cool. Some brief showers, referred to as the early or autumn rains, occur during this time. A few weeks later the main rainy season begins, lasting from December through February. Snow in Israel is a rare occurrence and even then is usually found only at higher elevations. By the end of February, the rainy season is finishing and the temperature is again quite warm. During March and April, the later or spring rains provide additional moisture for the crops.

By the end of April or early May, the weather again becomes quite hot, and there is usually no rainfall again until September. However, fairly regular dew keeps plants from totally drying up. The weather during the summer becomes very hot, especially in August.

Because there are no major rivers in the West Bank, rainfall is very important. Without adequate rain, the crops will not grow. In Deuteronomy 11:10-11 we read:

> The land you are entering...is not like the land of Egypt, from which you have come, where you planted your seed and irrigated it by foot as in a vegetable garden. But the land you are crossing the Jordan to take possession of is a land of mountains and valleys that drinks rain from heaven.

Since the Lord is the one who provides rain, the dry, hot weather made Israel very dependent on God, and he promised to provide the needed rain:

> It is a land the LORD your God cares for; the eyes of the LORD your God are continually on it from the beginning of the year to its end. So if you faithfully obey the commands I am giving you today—to love the LORD your God and to serve him with all your heart and with all your soul—then I will send rain on your land in its season, both autumn rains and spring rains, so that you may gather in your grain, new wine and oil. (Deuteronomy 11:12-14)

However, the rainy season was not the only vitally important source of moisture for crops. Even during droughts some crops survived and bore fruit because of the heavy dew that regularly appears even in the hot season. This fact helps us understand the terrible curse which God sent Elijah to speak to King Ahab: "As the LORD, the God of Israel, lives, whom I serve, there will be neither dew nor rain in the next few years except at my word" (1 Kings 17:1).

In order for the Israelites to have an abundant harvest, however, the Lord needed to provide both good autumn and spring rains. Joel uses this picture in one of his prophecies:

> He [God] sends you abundant showers, both autumn and spring rains, as before. Your threshing floors will be filled with grain; the vats will overflow with new wine and oil. (Joel 2:23-24)

Finally, we should also note that the winds were a very important factor in Israel's weather. During the dry, hot summer season, the wind usually came from the west or northwest. Coming off the Mediterranean Sea, the moist and cooling wind helped to bring comfort from the intense heat. Often during other times of the year, a hot, dry wind blew from the south or the east. This kind of wind with its scorching heat could quickly wither and kill the grass (Psalm 103:16; Ezekiel 19:12).

C. CROPS AND FESTIVALS

In our next lesson, we will learn about ancient farming methods. Here we will only touch on the main points of the growing seasons and how they fit in with the religious festivals of the Jews.

In the middle of Chart 6, you can see a list of the main events of the agricultural year in Israel. During the month of October the fields were plowed as the early rains began to soften up the soil. Then the Israelites did their main planting of grain—wheat and barley. These crops would be ready for harvest in late spring, from March through May. The grain was threshed over the summer. Israel's other main crops were olives and grapes. Grape vines were pruned about December, blossomed in May, and were harvested in September. Olives were harvested at about the same time.

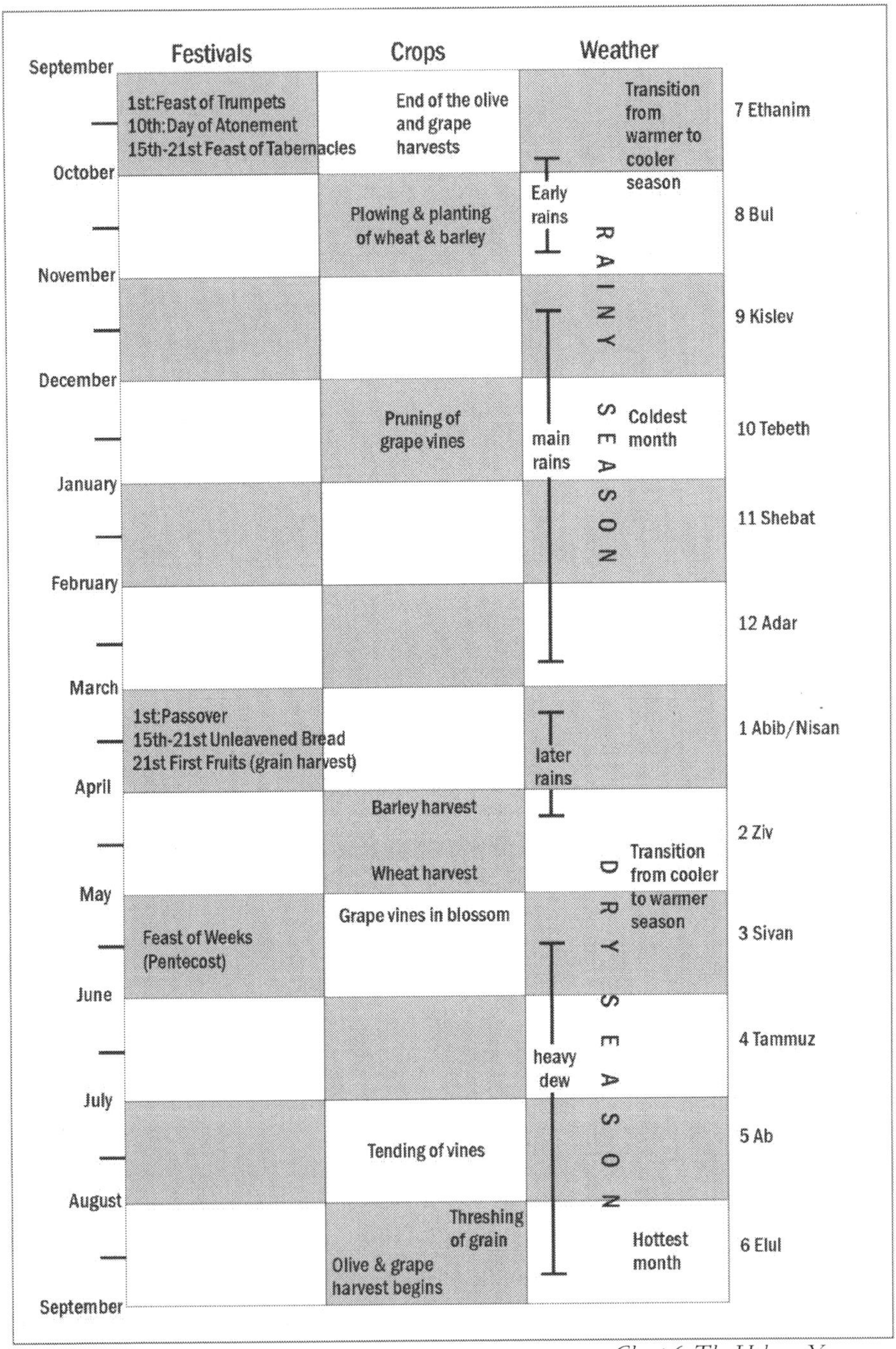

Chart 6. The Hebrew Year

The left column of the chart shows Israel's major religious feasts or festivals (these two terms are used interchangeably). On the 14th day of the first month (Abib or Nisan) was the great Festival of Passover. It was immediately followed by the Feast of Unleavened Bread, which lasted from the 15th to the 21st of the same month (Leviticus 23:4-8). The Festival of First Fruits was on the last day of the Festival of Unleavened Bread. On that day the Israelites brought the first fruits of the grain harvest as a special offering to the Lord, for the grain harvest was just beginning (Leviticus 23:9-14).

Seven weeks or 50 days later Israel celebrated the Feast of Weeks (Leviticus 23:15-21). By New Testament times, it was referred to as Pentecost (the Greek word for "fifty," since it came 50 days after Passover, Acts 2:1). This festival celebrated the end of the grain harvest, and it fell towards the end of the third month, Sivan (early June). It gave the Israelites a special opportunity to thank God for the grain harvest. It was also a time when the people recalled God's giving of the Law to Moses.

The final set of festivals was in Ethanim, the seventh month (September or October). On the first of the month was the Feast of Trumpets. It celebrated the beginning of the agricultural year (Leviticus 23:23-25). On the 10th day of the same month was the Great Day of Atonement (Yom Kippur), when the high priest entered the Most Holy Place to make atonement for the people (Leviticus 23:26-32). Then the Feast of Tabernacles followed, which lasted from the 15th to the 21st. During this week, the people lived in tents to remind themselves of the Exodus from Egypt (Leviticus 23:33-43). This last feast was also called the Feast of Ingathering because it also celebrated the final harvest of grapes and olives (Exodus 23:16). Once again, the children of Israel were to give thanks to the Lord for all the blessings of fields and flock which he had bestowed on them.

Three times each year all Israelite men were to go to the Tent of Meeting, or, later, to the temple, to worship the Lord and bring him their offerings. These three great festivals were 1) the Passover and the Feast of Unleavened Bread, 2) the Festival of Weeks, and 3) the Feast of Tabernacles (Deuteronomy 16:16-17). All three of these special worship occasions were times of thanksgiving to God for granting Israel a successful harvest.

LESSON 15:
GRAIN, GRAPES, AND OLIVE OIL

(God is) bringing forth food from the earth: Wine that gladdens the heart of man, oil to make his face shine, and bread that sustains his heart. (Psalm 104:15-16)

These words of the Psalmist picture a time of happiness and joy in Israel. Most people cannot be happy if they are hungry. For Israel, the items mentioned above suggested the abundance of God's earthly blessings for his people—wine made from grapes, oil from olives, and bread made from grain. These were the three main items in ancient Israel's diet and economy, and they were also the symbols of their prosperity. In this chapter we will examine how these crops were grown and the importance of their use.

By the end of this lesson you should be able to:
* *Briefly describe the nine main steps in growing and preparing grain;*
* *Describe the equipment needed to grow and prepare grain;*
* *Name the two main types of grain grown in ancient Israel;*
* *Describe an ancient vineyard and winepress;*
* *Give a description of an olive press; and*
* *Name five uses for olive oil in Bible times.*

A. GRAINS FOR MAKING BREAD

The two most important grains grown in Israel were wheat and barley. Both of these were essential for making bread and other foods. There were nine main steps the people of Israel followed in growing and preparing these grains.

1) Sowing or planting the seeds of grain began when the early rains came and softened the ground. The seed was kept in a basket or bag, and the sower would walk up and down his field, scattering the seed by hand. Jesus describes this process in the Parable of the Sower (Matthew 13:3-8).

2) Plowing the soil was the next step. Usually, another man followed right behind the sower immediately after the sowing. Ancient plows were usually made of wood and tipped with metal. They dug into the earth but did not turn it over like modern plows. The plowshare was attached to a wooden beam and then a yoke. The plow was usually pulled by a pair of oxen, although other animals were also used. Poor farmers might have just a single animal to pull their plow, and some might even have to pull it themselves. Plowing in the rocky soil of Palestine was difficult work, so over time, the yoke became the symbol of a heavy burden (Jeremiah 28:11; Matthew 11:29-30). Plowing was also hard work for the driver. He had to press down hard to get the plow to cut deeply, and he had to keep his animals going in a straight line. That is why it would be foolish for a farmer to look backwards while plowing (Luke 9:62). To keep the animals at work and pulling in the proper direction, the driver carried a goad, a long wooden stick, sometimes tipped with iron, to prod or beat the animal (1 Samuel 13:21; Acts 26:14). The judge Shamgar used such a goad to kill 600 Philistines (Judges 3:31).

3) Growing took several months and was a dangerous time for the grain. Birds might come to eat the seed, or weeds might hinder its growth (Matthew 13:4, 25). Fire might destroy the field of growing grain (Exodus 22:6). Swarms of locusts might eat the crop (Deuteronomy 28:42). Thieves might steal the grain (Judges 6:3). And a shortage of rain could dry up part or all of the crop.

4) Harvesting took place in April and May. The grain was usually harvested by hand with a sickle (Joel 3:13). Sickles were often made from the jawbones of animals, with pieces of sharpened flint fixed in them to form a sharp cutting edge. Samson didn't need the help of sharpened flint; he was able to cut down a thousand Philistines with a similar sickle made from the fresh jawbone of a donkey (Judges 15:15 ff.). After the grain was cut, it was gathered into bundles or sheaves (Genesis 37:7) and then taken from the field to the threshing area. The Law of Moses allowed hungry or poor passers-by to harvest some of the grain by hand while it was

ripening in the field. However, they were forbidden to take large amounts by cutting the grain with a sickle (Deuteronomy 23:25; Luke 6:1-2). Also, harvesters were always supposed to leave a little grain around the edges of the field so that the poor could gather or glean it (Leviticus 23:22; Ruth 2:2-3).

5) Threshing was the next step. This process involved crushing the stalks of grain, thus separating the individual grains from the stalks. Workers did this on a threshing floor, a flat open area about ten yards square with its earth packed down and cleaned. Threshing floors were usually located at the top of a hill so the wind could blow away the chaff as the grain was being winnowed. Araunah had such a threshing floor on top of Mt. Moriah (Mt. Zion). David bought the threshing floor and later Solomon built the Lord's temple on that land (2 Samuel 24:18). To thresh the grain, the bundles were undone and the stalks of grain were spread out on the threshing floor. The grain could be threshed or crushed in several different ways. Sometimes oxen were driven over the grain to crush it with their hoofs; at other times a threshing cart or sledge made up of beams was pulled by the oxen across the grain (Isaiah 28:27-28; 2 Samuel 24:22). Sharp stones or metal were at times set into the beams to make the sledge cut better (Isaiah 41:15). People might also stand on the beams to increase the weight of the sledge. If small amounts of grain were threshed, or there was no threshing floor available, a person could hit the stalks repeatedly with a flail, a stick with several other sticks or leather straps attached to form a type of whip. This is probably how Ruth and Gideon threshed grain (Ruth 2:17; Judges 6:11).

6) Winnowing followed next. It could be done only after the grain was thoroughly crushed and separated from its stalks. Winnowing was usually done in the evening when the winds arose. Especially with small amounts, grain could be winnowed by placing it on a garment and then throwing it into the air. Usually, however, a winnowing fork (like a pitchfork) or a winnowing fan was used. The worker threw the grain high into the air with the fork. The wind then carried away the stalks and husks, but the heavier

kernels of grain fell back to the ground. The stalks and husks were referred to as the chaff or stubble, and it was either allowed to blow away (Job 21:18) or was burned (Matthew 3:12; Isaiah 47:14). God used the separating of grain from chaff by winnowing as a picture of his judgment on his unfaithful people (Jeremiah 15:7; Matthew 3:12).

7) Sifting came next. Sifting was necessary because, even after winnowing, small pieces of dirt, weeds, or other impurities sometimes still remained in the grain. Sieves were used to sift the grain before it was stored or sold (Amos 9:9; Luke 22:31).

8) Storage was necessary to preserve the grain for future use. Large amounts of grain were usually stored in storage pits or barns (Matthew 13:30), while smaller amounts were most often stored in pottery jars (Ezekiel 4:9). These could be sealed to keep out insects and animals.

9) Grinding usually did not take place until it was time to bake the grain into bread. Small amounts of grain could be ground into flour using a mortar and pestle (Proverbs 27:22). For large amounts, however, two heavy millstones were used (Deuteronomy 24:6). The grain would be poured into a hole in the top millstone which was then rotated. The grain would fall between the two stones and be crushed as the top stone turned. It was heavy work for men to turn the stones (Lamentations 5:13), so often oxen were used to turn the stones (1 Corinthians 9:9). The stones were so large and heavy that they could crack a skull (Judges 9:53). If a person had such a millstone hung about his neck and was then thrown into the sea, it would have meant certain death—a picture of God's terrible judgment which Jesus used in his sermons (Matthew 18:6). After grinding, the flour was then put in a jar and was ready for making into breads and cakes (1 Kings 17:12).

B. THE GROWING OF GRAPES AND THE MAKING OF WINE

> I will sing for the one I love a song about his vineyard: My loved one had a vineyard on a fertile hillside. He dug it up and cleared it of stones and planted it with the choicest vines. He built a watchtower in it and cut out a winepress as well. Then he looked for a crop of good grapes, but it yielded only bad fruit. (Isaiah 5:1-2)

Vineyards could be found throughout Israel. Those planted on hillsides were especially good since they received extra sunlight which made the grapes extra sweet. In planting a vineyard, the land for the vines was carefully cleared. Farmers used hoes to prepare the ground (Isaiah 7:25). They then planted good young shoots in the prepared ground. Often the owners built a wall around the vineyard to protect it from animals and thieves. Watchtowers were also built and a watchman kept guard while the grapes were ripening. He would chase away birds and animals as well as watch for thieves. The farmer also had to be on guard against worms and other pests (Deuteronomy 28:39). Frequently, a winepress was built right in the vineyard so that the grapes could be processed into wine quickly.

Although there was a good deal of work in establishing and caring for a vineyard, the profits could also be great. When a man owned a vineyard and took care of it, he expected to get a good harvest. If the plants produced nothing, bad fruit or a poor harvest, he had a right to be angry (Isaiah 5:4-5; John 15:1-17). Wealthy landowners would build vineyards and then have their servants take care of them or would rent them out (Matthew 21:33-34).

One of the benefits of growing grapes is that they can be put to many uses. They can be dried into raisins (1 Samuel 25:18) or eaten fresh (Deuteronomy 23:24), and the juice could be boiled to make a thick sweet syrup or "grape honey." Some of the Old Testament references to honey might be to this grape honey (Genesis 43:11; Ezekiel 27:17). But grapes were most often used to make wine. The seriousness of the Nazarite vow was made clear by the fact that not only was a Nazarite to drink no wine, but he wasn't to eat any grape products at all (Numbers 6:4).

The process of making wine began at the winepress. Winepresses were open areas made up of two compartments. The grapes were put in the first section and squeezed or stomped on to obtain their juice. This picture was often used to portray God's judgment upon sinners (Revelation 19:15). The people often sang as they crushed the grapes since this was a festive time (Jeremiah 48:33). As the juice began flowing, it ran out of the first compartment through a small hole and entered the second and lower compartment of the press. The juice was then removed and placed in animal skins (Joshua 9:4). The juice was fermented by the natural chemicals in the skins and soon became wine. New wine had to be placed into new skins which were still soft and pliable and could expand as this took place. Old skins that had become hard would burst under the pressure of the gasses formed during the fermentation process (Matthew 9:17). The wine could be put into jars for storage or shipment.

C. OLIVES FOR MAKING OIL

The olive was the third of the three most important crops of ancient Israel. Olive trees, like grapevines, could grow even in rocky soil (Deuteronomy 32:13). Yet it took seven years after a tree was planted or grafted before it bore its first fruit. It would be another seven years before it would be fully grown and bear a complete harvest. Thus a land needed long periods of peace for its olives to grow and prosper, and the olive branch became a symbol for peace. Because the trees took so long to mature, destroying olive groves was a much more ruthless form of war than merely burning vineyards and grain fields (Judges 15:5).

Once grown, olive trees could continue to produce fruit for hundreds of years. New shoots sprang up around the older olive plants to carry on even longer (Psalm 128:3). However, olive trees that grow wild do not produce edible fruit. Such plants need to be pruned and a good branch grafted in before they will produce good olives. St. Paul uses this picture to describe the growth of the church (Romans 11:23-24).

Olives were harvested by beating the tree with brooms or branches (Isaiah 17:6). The Israelites sometimes dried and ate the fruit, and olive wood was used for building or carving statues (1 Kings 6:23, 31-33). But the most valuable product of the olive tree was the oil squeezed from the olives. Olives are much firmer than grapes, so an olive press was needed to squeeze the oil from them (Micah 6:15). Since olive oil was harder to extract, the presses had to use levers, torque, and greater weight to squeeze out every last drop of the precious oil.

Making olive oil was hard work, but it was well worth the effort. The peoples of the ancient world used olive oil in a great number of ways. They used it as a kind of butter for eating, or for cooking and baking (Leviticus 2:4). They used it as fuel in their lamps (Matthew 25:3), as an ointment for moisturizing skin, and as a medicine (Isaiah 1:6). They also used it to anoint the men whom God appointed to do his work—the prophet (1 Kings 19:16), the priest (Leviticus 8:12), and the king (1 Samuel 16:13).

Thus, when the Israelites had a good grain harvest and a bountiful crop of grapes and olives, they could truly see this as a blessing of the Lord and could look forward to a year of prosperity.

LESSON 16:
THE TIME OF THE JUDGES

After the conquest of the Promised Land, Joshua continued to rule over the people of Israel until his death. Then, for more than 300 years Israel was ruled by a series of judges. In this lesson we will study these men.

At the end of this lesson you should be able to:
* *Briefly describe the pattern of events which repeatedly took place in the time of the judges;*
* *Give the number of the judges, and the dates for this period;*
* *Describe the weapons of Shamgar, Gideon, and Sampson;*
* *Describe where and against whom Ehud, Barak, Gideon, and Samson fought their battles.*

A. THE TIME OF THE JUDGES

The period of the judges was a time of troubles for Israel. Invasions by their enemies, a decline in religious and moral values, crop failures, and military defeats at the hands of some of the nations around them all took place during this time. Why did this happen to Israel? Scripture answers this way:

> After that whole generation had been gathered to their fathers, another generation grew up, who knew neither the LORD nor what he had done for Israel. Then the Israelites did evil in the eyes of the LORD and served the Baals. They forsook the LORD, the God of their fathers, who had brought them out of Egypt. They followed and worshiped various gods of the peoples around them. They provoked the LORD to anger because they forsook him and served Baal and the Ashtoreths. In his anger against Israel the LORD handed them over to raiders who plundered them. He sold them to their enemies all around, who they were no longer able to resist. Whenever Israel went out to fight, the hand of the LORD was against them to defeat them, just as he had sworn to them. They were in great distress. (Judges 2:10-15)

To address this situation, the Lord sent the men called judges. The judges, however, did not completely solve the problem:

> Then the LORD raised up judges, who saved them out of the hands of these raiders. Yet they would not listen to their judges but prostituted themselves to other gods and worshiped them. Unlike their fathers, they quickly turned from the way in which their fathers had walked, the way of obedience to the LORD's commands. Whenever the LORD raised up a judge for them, he was with the judge and saved them out of the hands of their enemies as long as the judge lived; for the LORD had compassion on them as they groaned under those who oppressed and afflicted them. But when the judge died, the people returned to ways even more corrupt than those of their fathers, following other gods and serving and worshiping them. They refused to give up their evil practices and stubborn ways. (Judges 2:16-19)

This pattern was repeated over and over again during the time of the judges. The Israelites worshiped false gods, God then sent an enemy to punish them, they repented, God sent a judge to save them, and then they soon began to worship false gods once again. Each time the cycle repeated itself, the Israelites fell deeper into idolatry. Even the moral strength of the judges themselves declined over time.

B. TWELVE JUDGES

The period of the judges began shortly after 1400 B.C. and lasted until about 1050 B.C. The Old Testament book of Judges mentions twelve different judges during this time, but there may have been more. For some the Bible gives us little more than their names; for others, it gives us longer accounts of their work in rescuing Israel. As you read about the judges in the following sections, look also at Map 15 which shows where the judges lived and where their enemies were located. The twelve most important judges are listed below by number (in the order in which they appear in the Bible). The same numbers are used on the map to indicate where each was living. The most important cities mentioned in the book of Judges are also shown on the map. 1 Samuel 1-8 mentions four additional judges (Eli, Samuel, Joel, and Abijah). Note also that these judges often served as regional leaders in Israel, and thus their times of service may have overlapped one another.

1) Othniel was the first judge mentioned in the Bible. He was a son-in-law of Caleb, from the tribe of Judah. After eight years of oppression by

Cushan-Rishathaim, a king of Aram (to the north of Palestine), God sent Othniel to drive the Arameans away. Forty years of peace followed (Judges 3:7-11).

2) Ehud was from the tribe of Benjamin. Eglon, a Moabite king, took possession of Jericho and from there oppressed the Israelites for 18 years with the help of the Ammonites and Amalekites. Ehud tricked Eglon into meeting him privately and then killed him with a knife he had concealed.
Ehud then led the Israelites to the Jordan and killed 10,000 of Eglon's fleeing men. Eighty years of peace followed (Judges 3:12-30).

3) Shamgar brought Israel temporary peace with Philistia by using an oxgoad to kill 600 Philistines (Judges 3:31).

4) Deborah was the next judge. She lived in a time when Israel was oppressed by the Canaanites, led by King Jabin of Hazor and his general Sisera. Deborah was a prophetess who lived between Ramah and Bethel. She guided the Israelite general Barak in defeating the Canaanites. Barak led his army down from Mt. Tabor and defeated Sisera's army near the Kishon River. Sisera himself was killed by the housewife Jael who drove a tent peg through his head while he was hiding in her house (Judges 4-5).

5) Gideon lived at a time when the Midianites and Amalekites had been stealing Israelite herds and food supplies for seven years. The Israelites had sought shelter in the mountains and in caves, but they were still not safe from these enemies. God had Gideon gather a small army of 300 men. They attacked the enemy in the Jezreel Valley at the Spring of Harod, near the Hill of Moreh. Using only trumpets, jars, and torches, the Israelites put the Midianites to flight. The Midianite kings and 120,000 soldiers died (Judges 6-8).

6) Tola was from the tribe of Issachar and led Israel for 23 years after Abimelech, the son of Gideon, had unsuccessfully tried to rule Israel (Judges 9-10:2).

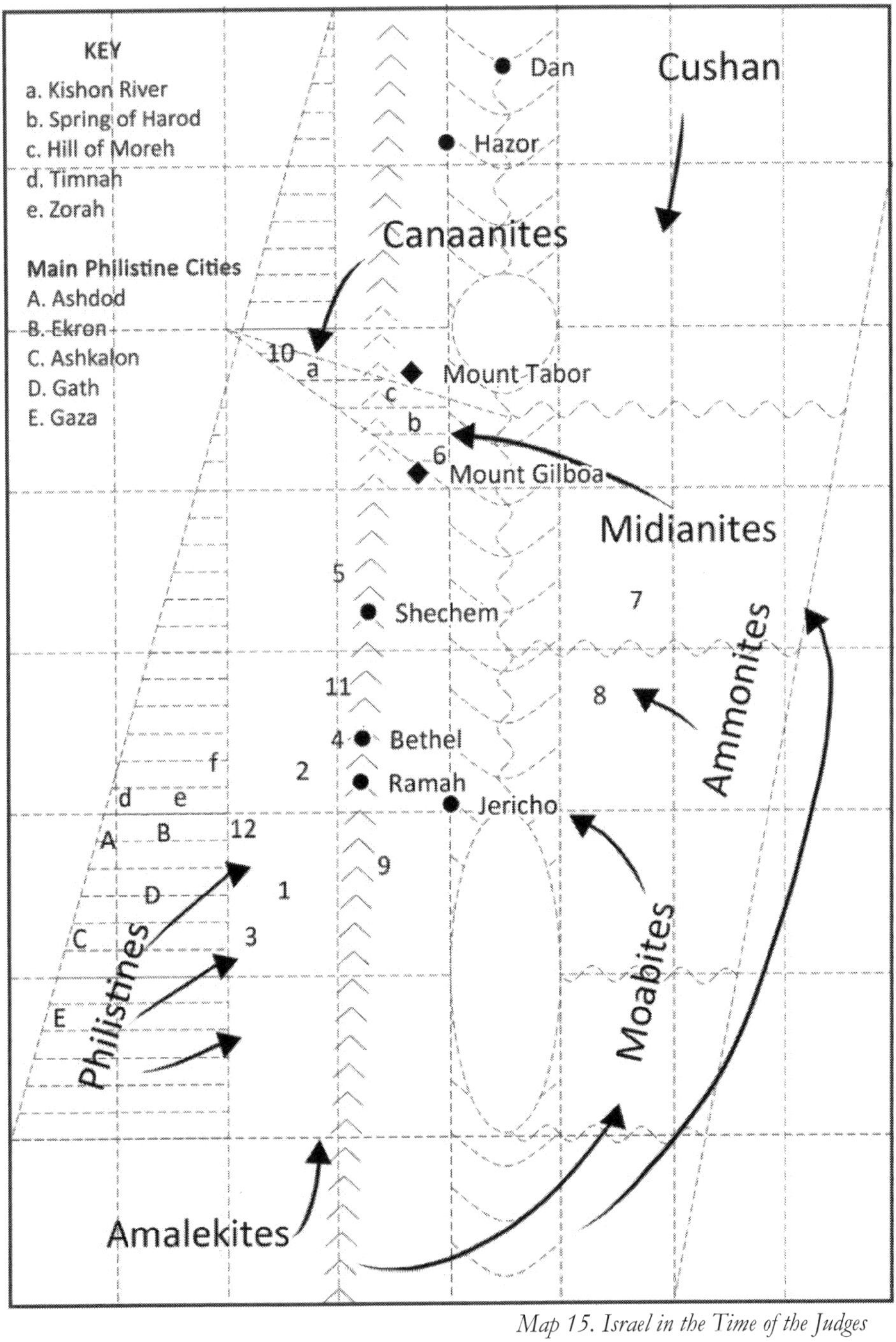

Map 15. Israel in the Time of the Judges

7) Jair ruled Israel for 22 years from Gilead on the East Bank (Judges 10:3-5).

8) Jephthah, who was also from Gilead, was asked by the Israelite elders to drive out the Philistines and Ammonites. These nations had oppressed Israel on both sides of the Jordan for 18 years. Jephthah and his men defeated the Ammonites and drove them from Gilead. He served as a judge for 6 years (Judges 10:6-12:7).

9) Ibzan of Bethlehem ruled Israel for seven years (Judges 12:8-10).

10) Elon, from the tribe of Zebulun, ruled Israel for ten years (Judges 12:11-12).

11) Abdon of Ephraim judged Israel for eight years (Judges 12:8-15).

12) Samson, from the tribe of Dan, is probably the most famous judge. Many of his adventures took place around the Valley of Sorek, a valley which leads from the seacoast up to Jerusalem along the southern edge of the corridor. During his time, the Philistines were troubling the Israelites. They had taken control of the shephelah and moved into the hills of Judea. Samson scored temporary victories over them, but they were not totally defeated until the time of King David (Judges 13-16).

Note that Judges 3:3 speaks of the five rulers of the Philistines. This refers to the rulers of the five main cities in Philistia, which are also shown on the map: A) Ashdod; B) Ashkelon; C) Gaza; D) Ekron; and E) Gath.

LESSON 17: THE UNITED KINGDOM OF ISRAEL

The period of the judges lasted for over 300 years. It was a period of struggles and difficulties for the people of Israel. It came to a sad end with Israel losing its entire spiritual and moral direction. Finally, the Israelites demanded that God give them a king so they could be like all the other nations around them. Initially, the kings brought Israel its greatest moments of glory along with a wonderful time of peace and prosperity. This period is called the United Kingdom. During this century (about 1050-950 B.C.), only three kings ruled, and it was remembered as Israel's Golden Age.

At the end of this lesson you should be able to:

* *Name the two leaders in the transition period between the judges and the United Kingdom;*
* *Relate the travels of the Ark of the Covenant as it was brought to Jerusalem;*
* *Give the dates for the United Kingdom and identify the extent of its borders at the height of its conquests; and*
* *Know the names of the first three kings of Israel, which tribe and city each came from and how long each ruled.*

A. ELI, SAMUEL, AND THE ARK OF THE COVENANT

After entering the Promised Land, Joshua set up God's Tent of Meeting at the city of Shiloh as a centrally located place for worship (Joshua 18:1). Toward the end of the time of the judges, in those years when there was no judge, the people of Israel turned to the high priest for guidance. The high priest Eli led Israel for 40 years, but he was a poor leader. He failed to discipline his two wicked sons who also served as priests. The result was both a military catastrophe and a loss of confidence in the leaders who governed Israel. The events surrounding the downfall of Eli and the activity of Samuel, the last of Israel's judges, form a transition between the period of the judges and that of the United Kingdom.

The Lord chose Samuel, a young boy serving in the temple under the guardianship of Eli, as Israel's next important leader. Samuel was ready to listen to and obey the Lord. Rather than judging Israel from one place, Samuel traveled a circuit throughout the land to better serve the people (1 Samuel 7:15-16). From his home in Ramah, Samuel traveled to Bethel, Gilgal, and Mizpah on his yearly circuit (see Map 16). Following God's directives (1 Samuel 8), Samuel first warned God's people of the future consequences of their request for a king, and then bowed to their demands and anointed Saul as the first king of Israel. After Saul proved disobedient to God and unworthy to rule, Samuel also anointed Israel's second king, David.

Before the Israelites had a king, the Tent of Meeting and the Ark of the Covenant were the center of Jewish religious life. The Ark of the Covenant was a very special sign of God's presence among his people, but it had become an object of superstition. Many people believed that the Ark had supernatural powers. For this reason, the Israelites took the Ark with them into battle thinking that it would bring them victory. But the Lord allowed the Philistines to capture the Ark in a battle near Aphek. News of this tragedy caused the death of Eli (1 Samuel 4).

But the Philistines also suffered from capturing the Ark. Instead of bringing them power over the Israelites as they had hoped, a plague broke out in every city where they took it (1 Samuel 5). Frustrated with the problems they faced and fearful of future catastrophes, the Philistines finally sent the Ark back to Israel. It first came to Bethshemesh where 70 men died because they dared to look inside it. Then, it was sent to Kiriath Jearim where it remained for 20 years (1 Samuel 6) until David brought it permanently to his new capital, Jerusalem (2 Samuel 6).

B. SAUL

When the people insisted on having a king, God told Samuel to crown Saul. He was from the tribe of Benjamin and the city of Gibeah. Initially, Saul was a part-time king and continued to work his farm. He began with a genuine desire to serve the needs of his people (1 Samuel 9-11). But after Saul destroyed the Ammonites who were attacking the city of Jabesh Gilead, he gave up his farm and became a full-time king for the remainder of his 42-year rule. During that time, Israel tried to drive the Philistines out of their territory. Saul was assisted by his son Jonathan (1 Samuel 14). The young David also fought for Saul. When the people began to praise David's bravery, Saul became jealous and began to hate and persecute David.

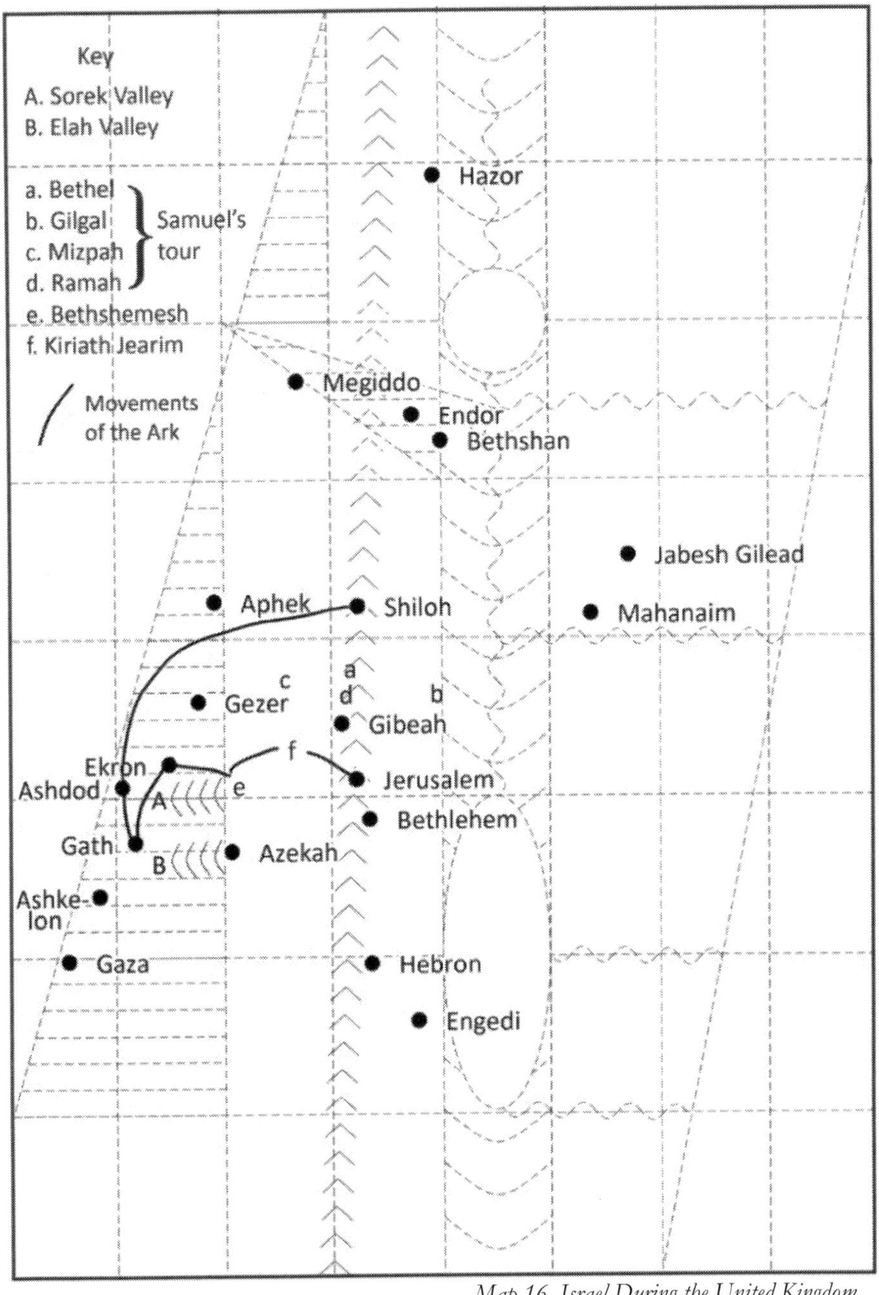

Map 16. Israel During the United Kingdom

As time went by, Saul disobeyed the Lord more openly. He conducted a sacrifice himself at Gilgal instead of waiting for Samuel (1 Samuel 13). He did not follow God's command to destroy all the men and property of the Amalekites after he had defeated them (1 Samuel 15). When the priests of Nob helped David with food, Saul had them all killed (1 Samuel 22). Because of Saul's cruelty and disobedience, the Lord took the kingship away from Saul and his family. While fighting the Philistines on Mt. Gilboa, Saul and Jonathan were killed. Their bodies were mutilated and hung on the city walls of Bethshan (1 Samuel 31). The people of Jabesh Gilead, remembering how Saul had once saved their city, secretly removed the bodies and buried them. In this way the Israelites' first king came to a bloody and shameful end.

C. DAVID

King David is the most well-known and best loved of all the kings of Israel. Under his leadership Israel enjoyed its largest period of expansion and became an influential country in the Middle East. Although a great warrior, David's most lasting work has been the nearly one hundred psalms which he wrote.

David was a shepherd from the tribe of Judah and the town of Bethlehem. He was the youngest of Jesse's eight sons. While still a boy, Samuel anointed David to replace Saul as Israel's next king, but for many years, David contentedly served under Saul until the Lord determined that his time had come. David's musical ability first brought him to Saul's attention, and he was appointed Saul's armor-bearer (1 Samuel 16). David also developed a lifelong friendship with Saul's son Jonathan, and even married Saul's daughter Michal. That God was richly blessing David was made clear when David killed the giant Goliath with his shepherd's sling (1 Samuel 17). This victory first stirred up Saul's hatred and jealousy of David.

During the following years, David spent much of his time as a fugitive, running from the jealous Saul. First, he fled to Samuel at Ramah. Then, pursued by Saul's army, he hid out at Nob (near the city of Jerusalem) where the priests supplied him with food to eat (1 Samuel 21). Finally, David was forced to leave the land of Israel and take refuge among the Philistines. Over time he gathered a small army of 400 men who remained loyal to him for many years. When he was forced to leave the region of the Philistines, David traveled south to Hebron and hid in the surrounding wilderness and rested for a time in the caves at Engedi on the shore of the Dead Sea. David had several opportunities to kill Saul, but each time he refused (1 Samuel 24). David waited for the Lord to punish Saul and give him the throne.

After the death of Saul and his son Jonathan, Saul's son Ishbosheth was proclaimed king over the northern part of Israel (2 Samuel 2). For seven and a half years, David ruled from Hebron, but only over the much smaller southern territory of Judah. After Ishbosheth was slain, David became king of all Israel as the Lord had promised. One of his first acts as king of all Israel was to move north and capture the fortified city of Jerusalem which he then made his capital (2 Samuel 5).

Jerusalem gradually became the center of a powerful kingdom. With God's guidance and blessing David greatly extended his kingdom to many times its original size. Its borders then reached from the Red Sea in the Sinai Peninsula and the edge of Egypt, to the great Euphrates River in the north. Israel never again controlled so much territory (2 Samuel 8). David spent almost his entire life fighting Israel's enemies, and he finally ended the power of the Philistines. Because he was a man of war, God did not allow David to build the Lord's temple in Jerusalem (1 Kings 5:3). That task fell to David's son Solomon, the great man of peace, who ushered in the next glorious era of the United Kingdom.

The Bible calls David "a man after God's own heart" (1 Samuel 13:14), that is, a man whose heart was centered in the Lord and his will and whom the Lord himself dearly loved. David was merciful to his enemies, full of thanks and praise for God, and filled with love for God and his fellow man. Nonetheless, David too was a sinner, as his adulterous affair with Bathsheba (2 Samuel 11) and his prideful taking of a census showed all too clearly (2 Samuel 24). He also experienced a broken heart through the wickedness and revolt of his own son Absalom. David had to flee to Mahanaim with his followers until Absalom had been killed (2 Samuel 15-18). After ruling for 40 years (7 in Judah and 33 over all Israel), David died at the age of 70 and was buried in his beloved city of Jerusalem (1 Kings 2).

D. SOLOMON

While David was a man of the heart and a man of war, his son Solomon was a man of the mind and a man of peace. He was the son of David and Bathsheba. As David lay dying, he heard that his eldest son

Adonijah was proclaiming himself as the new king. David then made it clear to all that Solomon was to be his successor (1 Kings 1:30).

Early in Solomon's reign, God offered to give him anything that his heart desired. Solomon requested the wisdom to govern God's people well (1 Kings 3). God granted his request, and Solomon is still known as one of the wisest men who ever lived. People came from vast distances (such as the Queen of Sheba who came from Southern Arabia, 1 Kings 10) to listen to him and speak with him. Solomon's wisdom led him to organize the country in a way that benefited the people of Israel. He built fortresses at Hazor, Megiddo, and Gezer in order to protect his country's borders. He operated a fleet of ships on the Red Sea (the only king of Israel to do so successfully), which brought in a wealth of goods from distant lands. He operated iron and copper mines in the Sinai Desert which also increased Israel's riches. During Solomon's reign, Israel was at the peak of its power and glory. And because of the magnificent palace, the splendid temple for the Lord, and the strong city wall, the capital city of Jerusalem became a major center of the Near Eastern world.

Although Solomon possessed great wisdom, he was not wise enough to avoid the three temptations Moses and Joshua warned against – making treaties with foreign nations, marrying heathens, and allowing the worship of foreign gods and idols (1 Kings 11:7-8). Solomon had a royal harem with 700 wives and 300 concubines (1 Kings 11). Some of these were foreign women, whom he married in order to cement alliances with other royal houses. In order to please his many foreign wives, Solomon granted them permission to build temples for their heathen gods on the hillsides around Jerusalem. As a result, despite his God-given wisdom, Solomon's sinful nature led him into idolatrous practices and the worship of false gods.

Solomon, like his father, ruled for 40 years, and God graciously allowed him to die while his empire was still at its peak. He was the author of three Old Testament books: Proverbs, the Song of Songs, and Ecclesiastes. Solomon's death not only brought about the loss of the United Kingdom's glory, but it also led to a period of spiritual decline within the nation of Israel.

LESSON 18:
OLD TESTAMENT JERUSALEM

As we noted earlier, Jerusalem became the capital of Israel during the time of the United Kingdom. Many of the Old Testament writers used the city of Jerusalem as a symbol or picture to describe the glories of heaven. In this lesson, we will learn more about the history and geography of Jerusalem during Old Testament times.

At the end of this lesson you should be able to:

* *List five names used in the Old Testament for the city of Jerusalem;*
* *Describe the important geographical features of Jerusalem by drawing a simple map;*
* *Describe the city during the times of both David and Solomon;*
* *Briefly recount how Jerusalem's two important valleys were used in ancient times;*
* *Explain what "Hezekiah's Tunnel" was.*

A. JERUSALEM BEFORE THE UNITED KINGDOM

Already 1000 years before David, a city was located on the site of Jerusalem. In Abraham's day the city was called Salem (Genesis 14:18). Melchizedek was the king of Salem during the time of Abraham. At the Lord's command, Abraham went to the area of Jerusalem to offer up Isaac on Mt. Moriah (Genesis 22:2). Records found in the ancient city of Ebla also mention the existence of the city already at that time. Documents from Egypt mention a city at Jerusalem in about 1900 B.C.

However, the Bible tells us little about Jerusalem until the time of the Exodus. The city was located just inside the southern border of Benjamin's territory and just north of Judah's territory. At that time the city was called Jebus, and its inhabitants were a tribe of people called the Jebusites. Joshua defeated the king of Jerusalem, Adoni-Zedek, along with other enemies of Israel on the day the sun stood still (Joshua 10:1). But although the Israelites defeated the king, they did not conquer the city and its inhabitants at that time (Joshua 15:63; Judges 1:21).

B. JERUSALEM UNDER DAVID

Jerusalem was a well-fortified city on a hill with deep valleys on two sides. Its central location and its fortified position made it an ideal choice for an important city. However, the Jebusites were still living in the city, so David and his men captured it (2 Samuel 5:6-10; 1 Chronicles 11:4-9). Jerusalem then began to be called the City of David. There, David established his new capital. He began by terracing the slopes of the hills around the city and building up their walls. He also rebuilt and fortified other sections of the city. It seems that Jerusalem in the time of David covered just the single hill called Ophel or Zion, so these names were also used to refer to the whole city (2 Chronicles 27:3; 2 Samuel 5:7).

David wanted to make this city Israel's permanent capital, so he did all he could to make it the center of Israelite life. He moved the Ark of the Covenant there so that Jerusalem became the center of Israel's worship life (1 Chronicles 15). Since God forbid him to build a temple as he had planned, David placed the Ark in a specially made tent and commanded that all normal sacrifices be held there (1 Chronicles 16:1-6). David also built a palace (2 Samuel 11:2), but Solomon soon replaced it with a much larger one. It was here that David died and was buried. However, since David was born and grew up in the city of Bethlehem, the Bible also refers to Bethlehem as the "city of David" (Luke 2:11).

C. JERUSALEM UNDER SOLOMON

During the plague which the Lord sent after David in sinful pride took a census of his troops, the prophet Gad told David to build an altar on the "threshing floor of Araunah the Jebusite" (2 Samuel 24:18). This threshing floor was located on Mt. Moriah, just north of the main part of Jerusalem in those days. Mt. Moriah is where Solomon later built God's temple (2 Chronicles 3:1).

David had already begun planning and gathering the supplies that would be needed for the temple's construction (1 Samuel 28-29), but the building of the temple still took Solomon seven years to finish. The temple's design was similar in its basic design to that of the Tent of Meeting. It was divided into two parts. The Most Holy Place was 30 feet (9 meters) square; the Holy Place was 30 feet (9 meters) by 60 feet (18 meters). The entire building was 45 feet (14 meters) high. In front was a small porch, 30 feet (9 meters) wide by 15 feet (4.5 meters) long with two bronze pillars named Jakin and Boaz (2 Chronicles 3:17). The temple had storerooms along its sides It was constructed of silver, gold, and cedar wood from Lebanon. The furnishings of the temple were similar to those of the Tent of Meeting, but even more richly ornamented. A courtyard for the priests surrounded the temple and an outer courtyard for the people lay further away. The temple was dedicated at the Feast of Tabernacles in the eleventh year of Solomon's reign (2 Chronicles 5-7; c. 960 B.C.).

In addition to the temple, Solomon also built his own palace, which took him 13 years (1 Kings 7:1). It was called the Palace of the Forest of Lebanon (1 Kings 7:2) because of all the cedar wood from Lebanon used to make it. The palace was 150 feet (46 meters) long and 75 feet (23 meters) wide. Solomon also built a Hall of Justice where he sat and judged court cases, as well as other palaces for himself and his many wives (1 Kings 7:7-8). These were probably built on Mt. Moriah, near the temple.

D. JERUSALEM AFTER SOLOMON

In the period of the Divided Kingdom, the city of Jerusalem continued to grow. A new area was added called the New Quarter (2 Kings 22:14; Zephaniah 1:10). An aqueduct was also built to connect the Gihon Spring on the east with the Lower Pool (Isaiah 7:3). When King Hezekiah (c. 700 B.C.) feared a siege by Assyria, he dug a tunnel through the rock beneath the city to bring water from the Gihon Spring outside the city walls to the pool of Siloam inside the city walls (2 Kings 20:20; 2 Chronicles 32:30). In the neighborhood around Jerusalem, especially in the Valley of Hinnom, Solomon had allowed heathen temples to be built and sacrifices made to idols. Other kings unfortunately followed his bad example.

E. THE GEOGRAPHY OF JERUSALEM

Map 17 shows Jerusalem and the surrounding area as it might have looked in Old Testament times. Notice that it was built around three valleys – the Hinnom, the Kidron, and the Cheesemakers' (which ran through the middle of the city).

The Valley of Hinnom (or Ben Hinnom, or Topheth) ran along the southern edge of the city. It was there that heathen people sacrificed their children to the god Molech (2 Kings 23:10; 2 Chronicles 33:6). This naturally gave the Hinnom Valley a bad reputation. It was used by many of Jerusalem's inhabitants as a place to dump their garbage. Because the garbage was often burned and smelly smoke was always rising from this valley, Hinnom (and its New Testament name, Gehenna) became a nickname for Hell.

The other major valley, just to the east of Jerusalem, was the Kidron Valley. It has a small stream flowing through it. In times of religious reform, believers burned the idols of the heathens and flung their ashes into the Kidron Valley and its stream (1 Kings 15:13; 2 Kings 23:12).

The smallest of the three valleys was the Cheesemakers' Valley. It is not mentioned in the Bible. It ran just west of the temple and the city on Mt. Zion. Note also the major hills. The early Jebusite fortress was built on Mt. Zion (2 Samuel 5:7). Solomon added Mt. Moriah to the city and built the temple there (2 Chronicles 3:1). Moriah eventually became known simply as the Temple Mount. To the east of the city, across the Kidron Valley, was the Mt. of Olives (Zechariah 14:4).

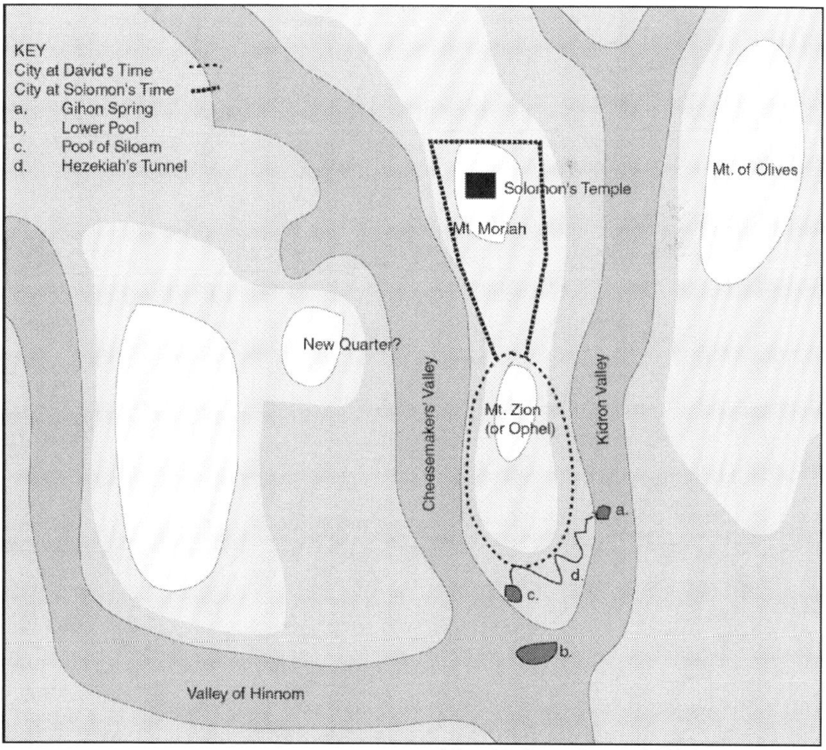

Map 17. Jerusalem in Old Testament Times

LESSON 19: ARCHAEOLOGY AND THE BIBLE

In our previous lessons, we have learned a great deal about the history, geography, and daily life of the Israelites during Bible times. Much of this information has not come from the Bible itself, but rather from the work of historians and archaeologists. In this lesson, we will learn what archaeology is, how it is studied, and how it can assist us in understanding the Bible.

At the end of this lesson you should be able to:

* *Define the words tell, excavate, potsherd, artifact, ostracon, and archaeology;*
* *Describe a tell and explain how it is excavated;*
* *Name some of the most common items found during an excavation;*
* *Name and give dates for the two earliest archaeological periods; and*
* *Briefly explain briefly what Bible students need to know about archaeology.*

A. WHAT IS ARCHAEOLOGY?

Archaeology is the study of the life and culture of the past through the objects that people and civilizations have left behind. The people who do this study are called archaeologists. Archaeologists study things like ancient cities, cemeteries, pottery, coins, weapons, clothing, jewelry, and whatever else has survived over the centuries. All such manmade objects from the past are called artifacts. Archaeologists use artifacts to learn about how people lived and about the history of ancient times.

Most buildings and other artifacts disappeared over long periods of time. Things of value were pillaged and re-used. Wind, rain, rust, and time eventually destroyed much of the rest. What was left behind was covered in dust or dirt which helped to preserve them. In order to find these again, archaeologists spend much of their time digging or excavating. Archaeologists dig in areas where people formerly lived, hoping to add to our knowledge of ancient civilizations.

In one of our early lessons we learned about the importance of water in the climate of the Near East. Because of the importance of water, cities were often built near springs, rivers, or by a well. Ancient cities were also often built on top of hills. This location provided a good view of approaching enemies and made the city easier to defend. Since Israel was also at the crossroads of ancient civilizations, the Israelites often had armies passing through their land and attacking their country. As a result, many of Israel's cities were destroyed repeatedly. But enemies were not the only cause of a city's destruction. Because ancient houses were made from sun-dried bricks or from stones, heavy rains or earthquakes could also cause the partial or complete destruction of cities. Yet because of the city's good location, a new city was simply erected on top of the ruins of the old one.

For all of these reasons, some cities in Palestine have been destroyed and rebuilt a dozen or more times. After each destruction, the ruins are flattened and the new city built on top. Thus the hills become higher and higher over time. A hill built up by layers of ruined cities is called a tell (also spelled *tel*). Tells have been around since ancient times but only relatively recently have archaeologists excavated and studied them for clues to their past (Joshua 11:13; Ezra 2:59).

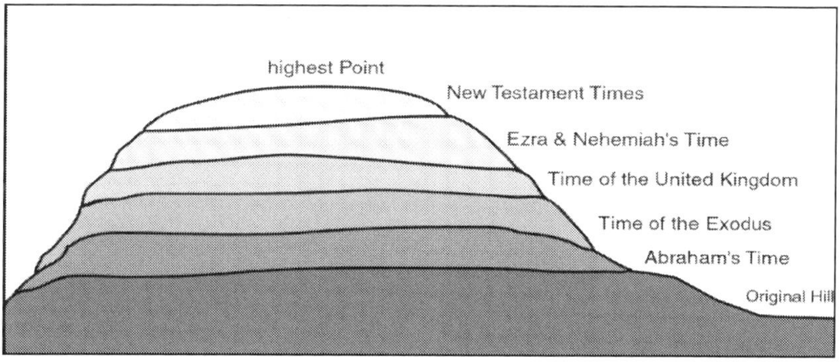

Chart 7. Example of a Tell

Chart 7 shows what a tell looks like. You can see that at the very bottom is the original hill on which the first city was built. In our example, you can see that the first city was built in the time of Abraham. When it was destroyed, it remained uninhabited until the time of the Exodus. That city was also destroyed, and a new city was built on top of it in the time of the United Kingdom. Such destruction and rebuilding continued to take place until New Testament times. In each layer of this tell are artifacts which give us information about the people who lived in this city at that time. Today, there are hundreds of tells all over Israel and the Near East. Some of them, like Jerusalem, have modern cities on them; others have been uninhabited for centuries. In these tells lie buried the remains of many important cities of ancient times.

B. DIGGING UP A TELL

In order to find out about the lives of the people who lived in ancient times, a tell must be dug up or excavated. This is the work of archaeologists, often in connection with universities. Before they can dig, however, they must gain permission from the government, raise money to pay for the project expenses, gather a group of other experts to assist, and organize a group of student volunteers or hire local laborers. Dozens of people are usually involved in digging up a particular tell.

Digging begins at the top and proceeds downwards. Usually a tell is too big to be excavated completely, so certain parts are carefully chosen for excavation. Digging often begins at the highest point, because important buildings regularly were built on the highest point (as is still done today). The most commonly found artifacts are pieces of broken pottery; these are called potsherds (or potshards). In ancient times, baked clay pottery was used for almost all the dishes, pots, pans, and bottles needed for everyday life. However, clay pots broke easily and could not be repaired, so they were usually just thrown aside. Thus in any excavation, thousands of pieces of broken pottery are found. Although they may seem useless, these potsherds are very helpful to the archaeologist. Since the shape, style, coloring, and decorations of the pottery changed over time, these broken pieces help archaeologists determine the time period when each layer of a tell was inhabited.

Archaeologists also hope to uncover the ruined walls or foundations of buildings. By carefully uncovering these remains, they can study the size and shape of ancient buildings and towns. As they continue to dig, they keep a careful watch for changes in the color of the soil or for burned ashes. These often indicate that a new level is being entered. Potsherds then help determine the dates of each new level. Work continues in this manner until the archaeologists reach the original level of habitation.

Archaeologists might discover any number of things as they excavate. One of the most useful is any form of writing. Writing may be found on stone, pieces of animal skin, sheets of papyrus, or on pieces of pottery. Pieces of pottery with writing on them are known as ostraca. In some parts of the Near East, whole libraries have been discovered, and the writings found provide invaluable aid in understanding the past.

C. ARCHAEOLOGICAL PERIODS

Archaeological digs have helped give us a more complete picture of mankind's history. Based on their findings, archaeologists have given names to the various periods of ancient history. Unlike the nine periods we learned in Lesson 3, these are not based on the Bible's account but on the artifacts found by the archaeologists. Thus, the period when people mainly used bronze for metal objects is named the "Bronze Age." In the following period, people learned to smelt iron, so this is called the "Iron Age." The later periods are named after the foreign cultures and powers that ruled over Israel after it lost its independence. Chart 8 lists the most important ages in Israel's history from an archaeologist's perspective:

The Bronze Age	3300 – 1200 B.C.
The Iron Age	1200 – 539 B.C.
The Persian Period	539 – 333 B.C.
The Hellenistic Period	333 – 63 B.C
The Roman Period	63 B.C. – A. D. 330

Chart 8. Main Archaeological Periods of Israel

These ages can be further subdivided. For instance, the Iron Age is often divided into Iron Age I (1200-1000), Iron Age IIA (1000-900), Iron Age IIB-C (925-585) and Iron Age III (585-539). Archaeologists do not always fully agree with one another, however, and they are constantly revising their opinions and the dates for these periods. Archaeology is not an exact science in all its aspects, so it does not give the same absolute truth that we get from the Holy Scriptures. Yet, archaeology does help us learn about the daily life and history of the people who lived in Bible times.

LESSON 20: TRADE, TRAVEL, AND TAXES

In Lesson 17, we learned about the United Kingdom, the great nation which God allowed David and Solomon to build up. During that century Israel enjoyed its greatest earthly glory and power and played a role in Middle Eastern politics and economy. In this lesson, we will learn about travel in Israel, the trade that passed through the land, and the taxes which were connected with trade.

At the end of this lesson you should be able to:

* *Name and describe the two main roads of ancient Israel which ran from north to south;*
* *Name four important exports of ancient Israel;*
* *Describe ancient taxes on trade goods and those who collected them; and*
* *Describe an ancient caravan and an inn.*

A. TRAVEL

In Lesson 12, we learned that when the Israelites conquered the West Bank of the Promised Land, they failed to conquer most of the coast of the Mediterranean Sea. As a result, Israel never had much to do with travel on the Mediterranean. In the time of Solomon, however, Israel controlled several ports on the Red Sea. Solomon used ships and sailors from in the Phoenician city of Tyre to carry on trade for him through these ports (1 Kings 10:11-12). Israel had few ships and sailors of its own. Small fishing boats sailed on the Sea of Galilee, but in general, travel in Israel was mainly by land.

To travel quickly and easily over land, people need good roads. This principle was as true in Bible times as it is today. Already in ancient times, there were important roads and paths used by long distance travelers. These roads were not covered with concrete, blacktop, or even gravel. They were often full of holes and bumps. The best ones had simple bridges to help the traveler cross swamps, streams, and rivers during the rainy seasons. Some also had a series of wells and rest houses along the road to provide food and shelter for the travelers.

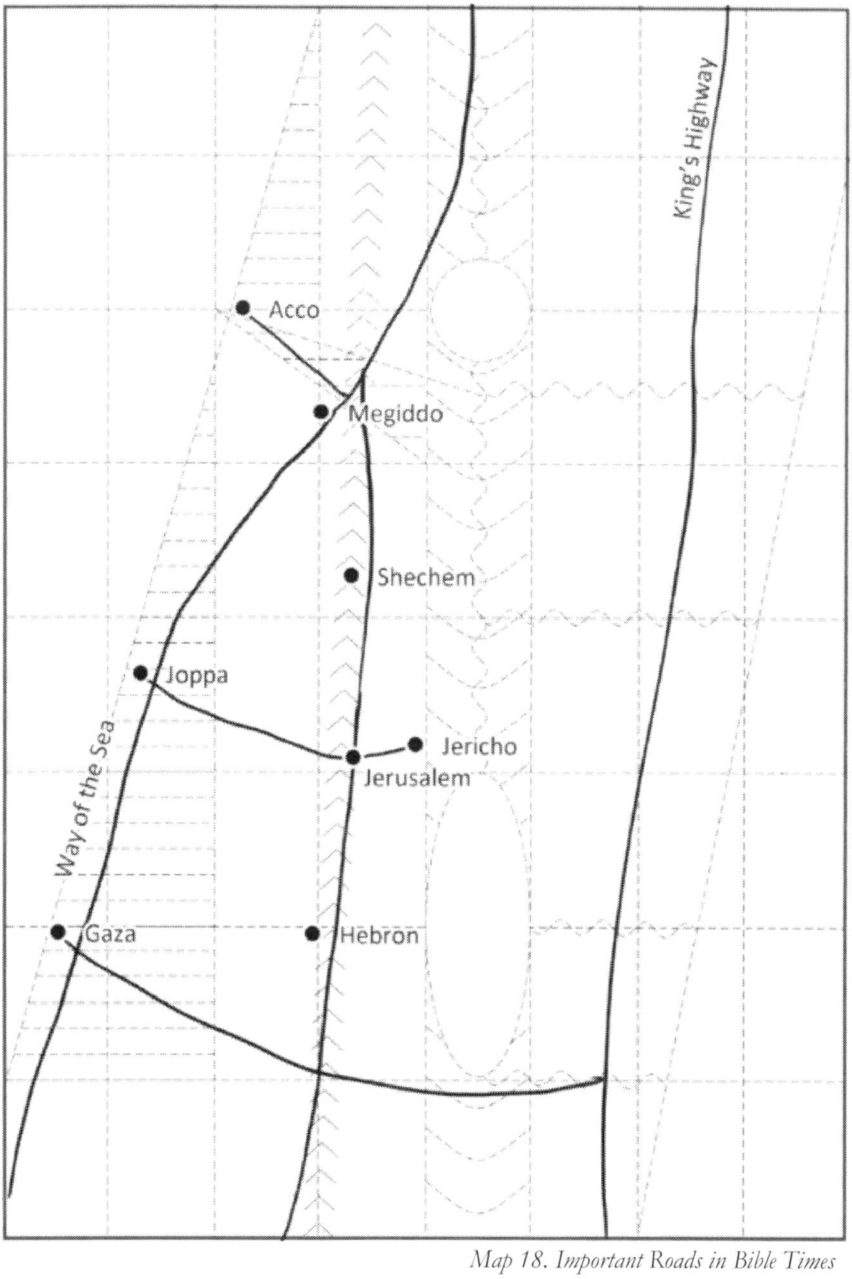

Map 18. Important Roads in Bible Times

Several important roads went from north to south through the land of Israel in Bible times. These roads are shown on Map 18. The most important road for traders ran through the Philistine Plain, crossed into the Jezreel Valley at Meggido, and then continued north through Galilee until it reached the city of Damascus. It was the main land trade route between Mesopotamia and Syria in the north and Egypt in the south. The southern part of this road was called the "Way of the Philistines" (Exodus 13:17). It was also called the "Way of the Sea" (Isaiah 9:1; Matthew 4:15) since much of it ran along the shore of the Mediterranean Sea.

The second most important road connecting Mesopotamia and Egypt was on the other side of the Jordan. This road was called the "King's Highway" (Numbers 20:17, 21:22). This road started near the Red Sea and ran between the inhabited areas and the desert all the way through Edom, Moab, Gilead, and Bashan; it continued on to Damascus. The Israelites wanted to travel along this road on their way to the Promised Land, but the Edomites and Moabites refused to allow them passage. These two roads served as the key passageways between north and south. Most long-distance travelers chose one of these two roads for their journey. They decided on one or the other based on the time of year, the stops they wished to make, safety concerns, and the political situation in the various areas through which they would travel.

A third road from north to south developed along the watershed, on the high ridge of hills down the center of Israel. Shechem, Jerusalem, and Hebron were important cities along this road. This road was especially important for travel within the country of Israel. Mary and Joseph may have taken this route on their trip to Bethlehem. A fourth possible route was to follow the Jordan River through the length of the Rift Valley.

Although the main trade routes ran from north to south in Israel, there were also several roads that ran from east to west. They connected the East Bank and the Rift Valley with the main ports on the Mediterranean Sea. The Jezreel Valley road started in the Rift Valley and followed the Jezreel Valley down to the Mediterranean port of Acco. Another road ran through the corridor, connecting Jericho and Jerusalem with the Mediterranean port of Joppa. Another road further south connected the lower Dead Sea area and the Negev with the important Mediterranean port of Gaza.

B. TRADE

Long before the time of Solomon, the civilizations of the ancient Near East engaged in international trade. Almost every country had its own specialties which were exported to other countries near and far. Ezekiel described the international trade situation in his day:

> Tarshish did business with you because of your great wealth of goods; they exchanged silver, iron, tin and lead for your merchandise. Greece, Tubal and Meshech traded with you; they exchanged slaves and articles of bronze for your wares. Men of Beth Togarmah exchanged work horses, war horses and mules for your merchandise. The men of Rhodes traded with you, and many coast lands were your customers; they paid you with ivory tusks and ebony. Aram did business with you because of your many products; they exchanged turquoise, purple fabric, embroidered work, fine linen, coral and rubies for your merchandise. Judah and Israel traded with you; they exchanged wheat from Minnith and confections, honey, oil and balm for your wares. Damascus, because of your many products and great wealth of goods, did business with you in wine from Helbon and wool from Zahar. Danites and Greeks from Uzal bought your merchandise; they exchanged wrought iron, cassia and calamus for your wares. Dedan traded in saddle blankets with you. Arabia and all the princes of Kedar were your customers; they did business with you in lambs, rams and goats. The merchants of Sheba and Raamah traded with you; for your merchandise they exchanged the finest of all kinds of spices and precious stones, and gold. Haran, Canneh and Eden and merchants of Sheba, Asshur and Kilmad traded with you. In your marketplace they traded with you beautiful garments, blue fabric, embroidered work and multicolored rugs and cords twisted and tightly knotted. The ships of Tarshish serve as carriers for your wares. You are filled with heavy cargo in the heart of the sea. (Ezekiel 27:12-25)

This passage shows how widespread international trade was around 600 B.C. Ezekiel states that all these goods were being traded with the Phoenicians of Tyre. Many of the same goods would also have passed through Israel.

During the time of Solomon, Israel took part in this trade more than ever before. In addition to exporting wheat and oil, Solomon probably exported a great amount of copper from his mines in the Sinai region. This copper was sent out from the Red Sea to Arabia and Africa where it was traded for gold (1 Kings 9:28) as well as for "silver, ivory, apes and baboons" (10:22). Solomon also imported chariots and horses from Egypt for his own army, but he resold some of the horses to Syria and Asia Minor (10:28-29).

Whenever merchandise was carried across borders in ancient times, governments levied taxes on those goods. This practice went back to earliest history, and it became an important source of revenue for countries like Israel. Solomon received so much money from other sources, however, that he hardly needed the "revenues from merchants and traders" (1 Kings 10:15). Usually merchants paid this tax at the border when they entered a country. This may have been Zacchaeus's job in the border town of Jericho in New Testament times (Luke 19:2). Levi, better known as the disciple Matthew, seems to have collected taxes along a smaller road inside the country (Luke 5:27). Even short trips of goods often required paying tolls or taxes.

Many tax collectors overcharged people, and for this reason tax collectors generally had a bad reputation. They were usually required to collect for the government a certain amount of tax each year from each region for which they were responsible. Any money they collected beyond the set amount was theirs to keep. Therefore, it was a great temptation for them to overcharge and pocket the extra wealth for themselves. In New Testament times, Jews also hated the local tax collectors because they were considered to be collaborators with the Roman government. Thus tax collectors were viewed as both cheaters and traitors who worked for the colonial powers, so they were despised and ignored by other members of Jewish society (Matthew 18:17).

Traveling in ancient times was almost always dangerous, as Jesus's story about the Good Samaritan illustrates (Luke 10:30 ff.). Wealthy travelers or traders with valuable goods were especially in danger. As a result, people often traveled in larger groups. These groups were called caravans. Important people always traveled in large caravans, as the Queen of Sheba did when she visited Solomon (1 Kings 10:2). The caravan usually followed the main roads, stopping for the night at an inn. Inns were usually quite primitive by our standards, but they did provide a roof over one's head and a floor to sleep on. Their walls sheltered and protected the travelers, their animals, and their cargo from robbers and wild animals. Inns may have existed along the main roads already in the time of the patriarchs (Genesis

42:27). It was to such an inn that the Good Samaritan brought the wounded Jew in Jesus's story (Luke 10:34).

In Genesis we hear about the Ishmaelites (Midianites), famous long-distance traders of the time (Genesis 37:28). Later, the Ammonites and then the Nabateans became well-known traders. Camels and donkeys were the animals most commonly used for long-distance transport of cargo, and sometimes travelers rode those animals as well. Wagons were also used at times (Genesis 45:21), but they were usually too difficult to take over long trips because of the rough roads.

International trade flourished in Bible times. This trade encouraged the development of good roads, and the taxes paid on trade helped the economy of Israel and other countries.

LESSON 21: THE DIVIDED KINGDOM

The prosperity and power of Israel during the United Kingdom came to a very sudden end. A revolt took place, and Israel became two separate kingdoms. The period which followed, known as the Divided Kingdom, was a time of frequent trouble. It was also a period of weakness for the divided nation. In this lesson we will examine what caused the division of Israel, the problems that confronted the two kingdoms both from within and without, and the final destruction of both kingdoms.

At the end of this lesson you should be able to:
* *Explain how and why Israel became divided;*
* *Name the capital of each of the kingdoms;*
* *Name the most famous battle site of this period;*
* *Give the dates for the destruction of each kingdom;*
* *Name which foreign power destroyed each kingdom; and*
* *Name the two religious centers of the northern kingdom.*

A. ISRAEL IS DIVIDED

Though Solomon was a very wise king, the kingdom of Israel was already falling apart by the time he died. His people were tired of being forced to pay heavy taxes and to labor on his building projects. He had lost the respect of others because he had allowed his foreign wives to build altars and temples for their gods. The territories of Edom and Damascus were also in revolt against Solomon's rule and would soon regain their independence. Solomon was also deeply in debt to the king of Tyre. Above all, God himself was ready to punish Solomon for the false worship which he tolerated within Israel.

God sent the prophet Ahijah to prophecy that God would leave Solomon's son only a small part of the United Kingdom of Israel; the rest would be given to a new ruler – Jeroboam. Jeroboam was from the tribe of Ephraim and had been working as a construction supervisor over Solomon's building program in Jerusalem (1 Kings 11:26-28). Through Ahijah, God informed Jeroboam, "If you do whatever I command you, and walk in my ways...I will be with you; and I will build you a dynasty as enduring as the one I built for David" (11:38).

Solomon tried to prevent God's will by ordering his men to kill Jeroboam. But Jeroboam was warned of Solomon's intentions and fled to Egypt. After Solomon's death, his son Rehoboam became Israel's new king. Rehoboam refused to lighten the heavy burden put on the people by forced labor and taxes, so the people of the north revolted. Jeroboam returned from Egypt and became their king. Jeroboam ruled the northern 10 tribes from his new capital at Shechem. Rehoboam lived in his capital of Jerusalem and ruled the region of Judah. He also added most of the corridor to his kingdom, putting a greater distance between his enemy to the north and his own capital, Jerusalem.

B. TWO NATIONS AND FOUR ENEMIES

It was about 930 B.C. when the nation of Israel was divided into the two kingdoms of Israel and Judah. It would not be reunited again for many centuries. Although related by blood, the two nations became fierce enemies. They allied themselves for short periods against more powerful enemies, but afterwards they always reverted to their former hostilities. Unfortunately, neither kingdom followed the advice God gave Jeroboam through Ahijah.

Neither Judah nor Israel ever totally gave up the worship of false gods. Not surprisingly, God did not allow either nation to enjoy the prosperity and happiness of former times.

Chart 9 summarizes the period of the Divided Kingdom. On the left are the main events of the Kingdom of Judah. This was the name given to the southern kingdom made up of two tribes of Israel and governed by the descendants of David. Despite occasional reformations, Judah always lapsed back into the worship of false gods. On the right are the main events of the Kingdom of Israel, the name given to Jeroboam's government over the 10 northern tribes. The Kingdom of Israel, plagued by heathen worship, rapidly declined in power.

Throughout the era of the Divided Kingdom, God frequently punished both countries for their false worship. Often, he did this by means of invasions from their enemies. There were four main powerful enemies that threatened the Divided Kingdom during this period:

JUDAH		YEAR BC	ISRAEL	
Rehoboam	had many wives, false worship increased, Pharaoh Shishak plundered the Temple	930	JEROBOAM	put up golden calves at Dan and Bethel. Invaded by Pharoah Shishak.
Abijah			Nadab	evil
ASA-	reformed Judah, burned the idols, defeated Egypt. Some false worship allowed	910	BAASHA	evil
			Elah	evil
			Zimri	evil
			Tibni	evil; fought against Omri
			Omri	very evil;new capital at Samaria
JEHOSHAPHAT	good king as his father, but still allowed some false worship.	870	AHAB	very evil; married Phoenician Jezebel; much Baal worship ; time of ELIJAH; fought Damascus & Assyria
Jehoram	evil		Ahaziah	evil
Ahaziah	evil		Jehoram	mostly evil
Athaliah	daughter of Ahab, started Baal worship, killed David's family.	840	JEHU	anointed by ELISHA, killed Ahab's family and Jezebl. Tried to wipe out Baal worship; very bloodthirsty; paid tribute to Assyria.
JOASH	favored true worship but didn't end false worship. Paid tribute to Damascus.			
		800	Jehoahaz	evil; subject to Damascus
AMAZIAH	favored true worship, but allowed false; conquered Edom again; paid tribute to Assyria		JEHOASH	evil; captured Amaziah of Judah and plundered the Temple; paid tribute to Assyria; recovered Gilead
			JEROBOAMII	evil; got back Moab & Amon
UZZIAH	(also called Azariah) favored true worship but allowed false; rebuilt Jerusalem walls & army.		Zechariah	evil
		760	Shallum, Menahem, Pekahiah, and Pekah	evil
Jotham	favored true worship, didn't listen to ISAIAH, asked for help from Assyria		HOSHEA	subject to Assyria, revolts in 724, Israel overrun by Assyria
			722	SAMARIA DESTROYED PEOPLE OF ISRAEL DEPORTED
Ahaz	evil	700		
HEZEKIAH	great reform of worship; refused Assyria(Sennacherib) tribute, Jerusalem saved by Angel.			
MANASSEH	started Baal worship again ; killed true worshippers, later in life reformed.			
Amon	evil	640		
JOSIAH	great reform (even in north); cleansed Temple, Passover begun; discovered Law; died in fight with Pharoah Necho at Megiddo			
Jehoahaz	evil			
Jehoiachim	time of JEREMIAH	600		
Jehoiachin	evil			
Zedekiah	rebeled against Babylon			
587	JERUSALEM DESTROYED PEOPLE OF JUDAH EXILED			

Chart 9. Summary of the Kingdoms of Israel and Judah

1) Egypt, though not as powerful as in Abraham's time, was still a major threat, and occasionally was strong enough to try to regain a place of power in the Near East (1 Kings 14:25; 2 Kings 18:21);
2) Damascus was the center of a powerful state in Syria at this time (1 Kings 20 ff.; 2 Kings 6 and 16);
3) Assyria was the most feared and most powerful Middle Eastern nation for most of the period from 900 to 600 B.C. Assyria brought about the final destruction of the northern kingdom of Israel (2 Kings 17-19); we will study Assyria and Babylon more fully in the next lesson; and
4) Babylon was a city-state within Assyria during much of this period. However, starting about 625 B.C., Babylon rose to greatness once again, eventually ending Assyria's control of the region. God used Babylon to punish and destroy Jerusalem and carry off many people from the Kingdom of Judah into exile (2 Kings 24-25).

At various times, God used each of these nations to punish Israel and Judah for their sinfulness and unbelief. God also used the Philistines, Edomites, Moabites, and Ammonites to discipline his disobedient people. These smaller tribes raided both Israel and Judah from time to time. Because of their heathen worship and cruelty, the Lord also eventually punished each of these other nations.

C. JERUSALEM AND SAMARIA

Take a look at Map 19. As noted earlier, the capital of the Kingdom of Judah remained at Jerusalem throughout this period. Since by the end of the United Kingdom the tribe of Simeon had been scattered throughout Judah (fulfilling Jacob's prophecy, Genesis 49:6-7), the southern kingdom did, in fact, consist of only one complete tribe, Judah. However, its first king Rehoboam extended its borders by taking control over much of the corridor to protect himself from the north. His successors tried to maintain that new border. Judah also tried to extend itself into the East Bank below the Dead Sea by attacking Edom.

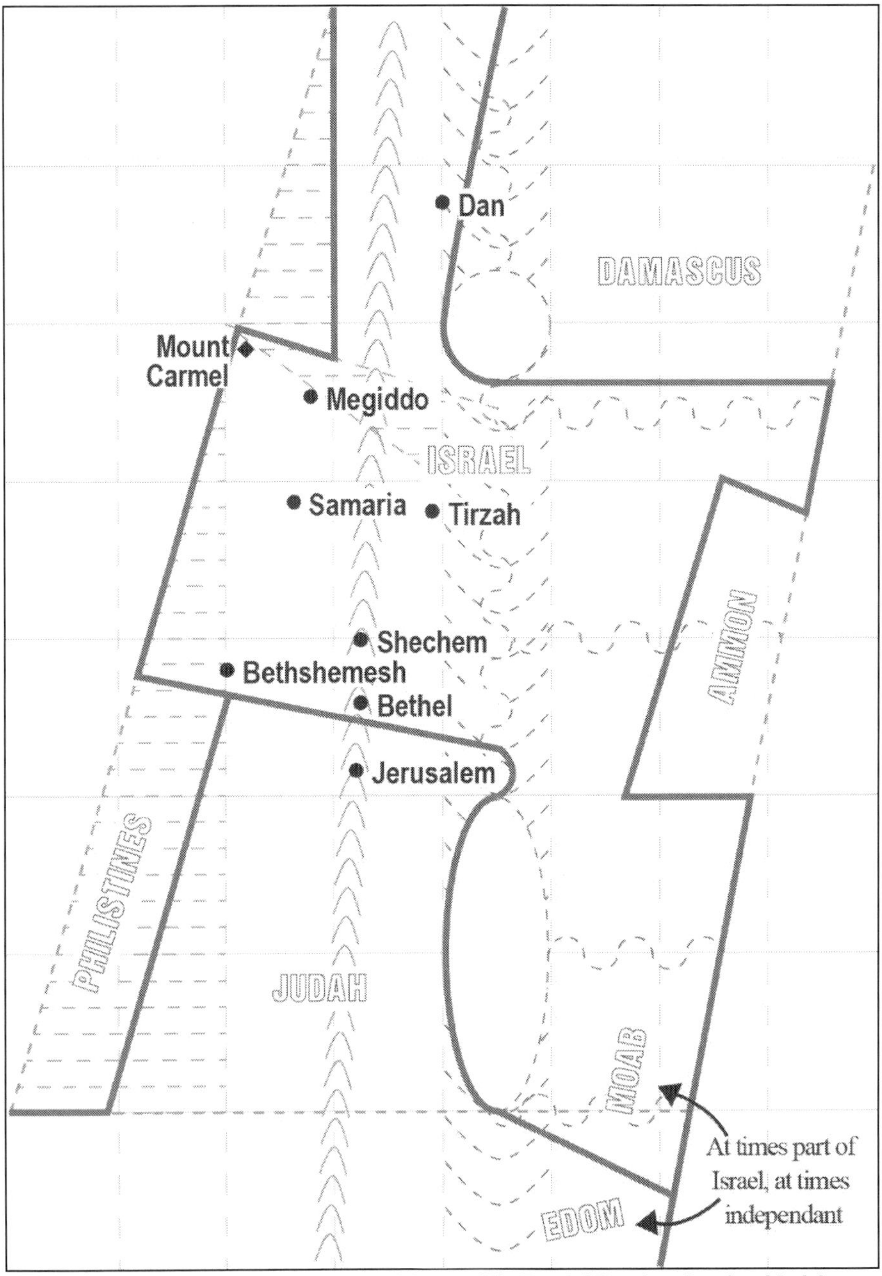

Map 19. The Divided Kingdom: Israel and Judah

Jeroboam, the first king of the northern kingdom, made his capital at Shechem. Since this city had been a famous religious and political center in the past (Joshua 24:1), it was chosen as the best substitute for the city of Jerusalem. Later, the capital was moved briefly to Tirzah (1 Kings 15:33), and then finally to Samaria (1 Kings 16:23-24). Samaria was both a strategic place and was quite centrally located, so it remained the capital of the northern kingdom. The cities of Dan and Bethel were also important religious centers for the northern kingdom. In each of those cities, Jeroboam set up a golden calf in order to give his people a place to worship, so they would not go to Jerusalem (1 Kings 12:28-33). Many true believers, however, moved to the southern kingdom to avoid this false worship (2 Chronicles 11:13-17).

One of the most strategically important cities during this period was Megiddo. It was located on the edge of the Jezreel Valley at the intersection of several major roads. When Pharaoh Shishak of Egypt marched against Jeroboam and Rehoboam, he made capturing Megiddo a high priority. King Ahaziah of Judah sought refuge there after being wounded (2 Kings 9:27), but he died shortly afterwards. In 733 B.C., the Assyrians captured it and made it their provincial capital for the region of Galilee. King Josiah of Judah chose the Pass of Megiddo as the only place he could hope to stop the advance of Pharaoh Necho's army; he succeeded, but he himself died in the battle (2 Kings 23:29). Because so many important battles were fought in this mountain pass, the book of Revelation used the term *Armageddon* (Hebrew for "mountain of Megiddo") as an earthly picture to describe the final battle of Christ and his church against Satan and his forces (Revelation 16:16).

The period of the Divided Kingdom was a time of confusion, war, and spiritual problems. Many important kings, prophets, and battles belong to these centuries. The many names and places often confuse Bible students. However, the history of the Divided Kingdom is important for showing the effects of sin and disobedience to God, as well as God's loving discipline and grace.

LESSON 22:
ASSYRIA AND BABYLON

In our last lesson, we learned that Assyria and Babylon were two of the most powerful countries during the time of the Divided Kingdom. These two nations destroyed the capital cities of the two Jewish kingdoms and carried off their people and wealth. In this lesson we will take a closer look at the nations of Assyria and Babylon.

At the end of this lesson you should be able to:
- * *Name the two periods in which Babylon was a great world power;*
- * *Name the period in which Assyria was a great world power;*
- * *Describe the Assyrian army and its "secret weapon;"*
- * *Describe the differences between Assyria and Babylon's deportation policies;*
- * *Briefly describe the Samaritans, their religion, and how the Jews treated them.*

A. THE MESOPOTAMIAN BACKGROUND

In Lesson 7, we learned that the Sumerians were a great civilization in Mesopotamia before the time of Abraham. Assyria and Babylon were located in that same area of the Fertile Crescent. Babylon is, in fact, the exact same area which was earlier called Sumer and Akkad. Assyria was farther north along the Tigris River, as Map 20 shows.

Shortly after the time of Abraham, from about 1900 to 1600 B.C., Babylon had become the capital city of a new empire. This period is called the Old Babylonian Period. Life and culture during this period were similar to that of ancient Sumer, but the rule of the famous king Hammurabi brought new stability. Hammurabi brought all Babylonia under his rule, organized it efficiently, ruled it well, and composed a set of laws for his people. Today, he is remembered most for his code of laws – the Code of Hammurabi.

From 1600 to 1150 B.C., the Kassite people controlled much of Mesopotamia. The Kassites came from the mountains northeast of Babylonia. They adopted most of the Sumerian and Babylonian culture, so the people's lives went on much the same as they had always done. Around

1350 B.C, the Assyrian people became a regional power with their capital city of Ashur. They remained dominant in the region until 600 B.C. Though Babylon briefly regained power now and then (1150-1050 B.C., for example), Assyria remained the most feared nation of the Mesopotamian region. From 1050-950 B.C., however, Assyria and Babylon both experienced a period of weakness. This weakness allowed Israel's United Kingdom to expand into a powerful nation.

As the period of the Divided Kingdom began, however, Assyrian power was again increasing. The Assyrian armies soon controlled most of Mesopotamia and large parts of Syria. Since the Assyrians had the most feared army in history, at times the kings of Israel and Judah asked Assyria for help against their other enemies (2 Kings 16:7-9). But King Ahab of Israel joined other kings of Palestine in trying to defeat Assyria. Ahab brought 2,000 chariots and 10,000 foot soldiers into the battle, but King Shalmaneser III of Assyria gained the victory (853 B.C.). Israel and Judah were forced to pay tribute to Assyria for most of the following century. Shalmaneser V later besieged Samaria for three years, because he was fed up with the constant rebellions (2 Kings 17:5-6). Sargon II finally overcame Samaria's resistance, and deported many of its people (722 B.C.).

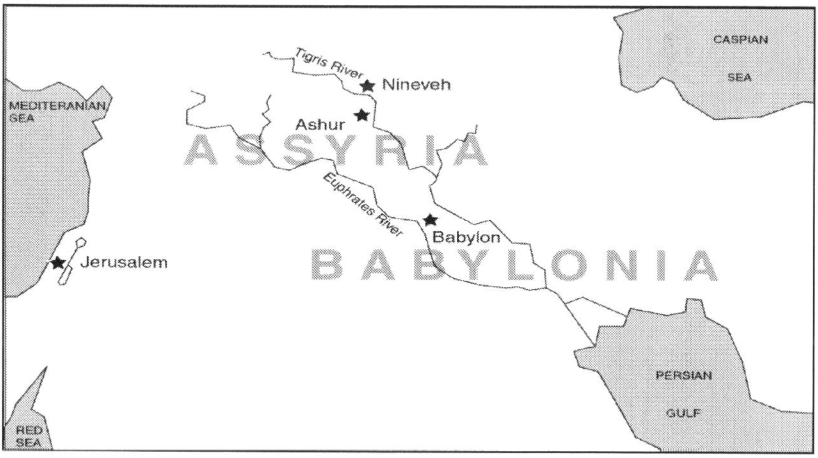

Map 20. Ancient Assyria and Babylonia

But Assyria's day of judgment also came. The prophet Zephaniah prophesied that God would "stretch out his hand against the north and destroy Assyria, leaving Nineveh utterly desolate" (2:13). And Nahum predicted that the Assyrian people would be "scattered on the mountains with no one to gather them" (3:18). About a century later, in 612 B.C., the Babylonians destroyed Nineveh, Assyria's capital. This period of Babylon's renewed power is called the Neo-Babylonian Empire. This time Babylon dominated the area from 625 until 539 B.C., when it was defeated by the Persian army of Cyrus. During the Neo-Babylonian period, Nebuchadnezzar II attacked and destroyed Jerusalem (597 B.C.; 2 Kings 24-25). This ended the Kingdom of Judah and began the period of the Babylonian captivity. Daniel lived and worked under the dynasties of Babylon's kings. Belshazzar, who witnessed the finger of God writing on his palace wall (Daniel 5:1 ff.), was Babylon's last king.

B. ARMIES AND WARS

Assyria is probably best remembered for its famous army. The Assyrian army was renowned for its use of chariots, its siege warfare, and the cruelty of its soldiers. Before this time, most armies were made up of just one type of soldier. Some countries had armies of foot soldiers; others used armies of cavalry or chariots. Some had archers skilled with bow and arrow; others had swordsmen; still others were known for their spearmen. The Assyrian army combined all of these elements of warfare. A group of scouts and spies went out ahead of the army to discover what the enemy was doing. The army itself had trained divisions of horsemen as well as archers and foot soldiers. It also contained religious officials and priests who offered sacrifices for victory. Scribes sent and received messages, engineers built bridges and roads, and other Assyrians carried and prepared the great amounts of food and other supplies needed by the army. In addition, a large group of back-up troops were brought in from the Assyrian provinces to support the regular army of Assyrians.

The "secret weapons" of the Assyrian army were its siege machines and battering rams. The battering rams were armored on the outside to keep enemy arrows and spears away from the soldiers who were operating inside. The soldiers rolled the machine up to a weak spot in the walls of the enemy city, and then a large ram in the front of the machine knocked a hole in the wall. The Assyrians also used moveable siege towers. Assyrian archers would be placed on top of them, and then the tower was moved into position so that the archers could pick off their enemies on the besieged city's walls. If the Assyrians were unsuccessful in knocking a hole in a city's walls, the army surrounded the city and cut off all food or water until the people starved and were forced to surrender. Using these tactics, the Assyrians besieged and conquered Samaria and then attacked Jerusalem. However, their attempted siege of Jerusalem was unsuccessful, in part because Hezekiah had quickly built a tunnel under Jerusalem in order to maintain a constant supply of water for the city (2 Kings 20:20).

C. TREATMENT OF REBELS

But the Assyrians were not just out to destroy other nations. If the people they conquered accepted their rule, the Assyrians were lenient with them. They didn't want to destroy other nations. They wanted the conquered nations to be prosperous so that they could pay tribute to Assyria. However, the Assyrians were extremely harsh in their treatment of rebels. Their cruelty served to keep the other nations under their control from joining any rebellion. The methods they used to end such opposition were to skin the rebels alive or skewer them on sharp posts.

In cases of repeated rebellion, the Assyrians took the educated people, leaders, craftsmen, and other important people from the rebelling city or nation and deported them to a new area. They resettled these leaders in small groups in many different cities and countries, so they would be absorbed into the native peoples living there. Then the Assyrians brought in small groups of rebellious people from other countries, or sometimes even loyal subjects, to replace the portion of the population which had been deported. In this way, the rebels lost contact with one another. Many lost their national identity. Thus, they were no longer able to plan and work together against Assyria.

The Assyrians used this tactic when they defeated Samaria and the northern Kingdom of Israel. "In the ninth year of Hoshea, the king of Assyria captured Samaria and deported the Israelites to Assyria. He settled them in Halah, in Gozan on the Habor River and in the towns of the Medes" (2 Kings 17:6; see also 15:29). Over the centuries there have been many legends claiming that the 10 northern tribes remained together wandering in exile, and that perhaps they might still be found or be able to be identified today. The Israelites who were taken into exile, however, were scattered among the other Mesopotamian peoples and absorbed into their populations. Thus, there are no "lost ten tribes of Israel" to be found.

As was their custom, the Assyrians imported other peoples to inhabit the land of the northern kingdom and replace the more educated and ruling classes of Israel society who had been deported. These foreigners came from areas throughout the Assyrian empire, and they intermarried with the Jewish peasants who had been left behind. Because these mixed-race people also mixed their religion with the Jewish religion, the Jews saw that they were unfaithful to the Lord. These are the people mentioned in the New Testament as the Samaritans, and they were regarded with hostility by all the following generations of Jews. 2 Kings 17:34 tells why:

> To this day these people persist in their former practices. They neither worship the Lord nor adhere to the decrees and ordinances, the laws and commands that the Lord gave the descendants of Jacob whom he named Israel.

Hatred between Samaritan and Jew continued into New Testament times. Like the Assyrians before them, when the Babylonians finally conquered Judah around 600 B.C., they too deported the vast majority of the educated Jews, while "the commander left behind some of the poorest people of the land to work the vineyards and fields" (2 Kings 25:12). Not too many were left, however, for the Bible tells us that most of the land remained unplowed and "enjoyed its Sabbath rests" (2 Chronicles 36:21). Those who were carried off were re-settled in large groups in Babylon. Unlike the Assyrian policy, the Babylonians did not scatter the captives among other peoples in order to make them lose their identity as a people. As a result, even after a 70-year captivity, large groups of Jews retained their Jewish identities and traditions in Babylon, and a portion of them returned to the Promised Land under Ezra and Nehemiah.

God had told the people of Israel through his prophet Isaiah, "Do not be afraid of the Assyrians.... Very soon...my wrath will be directed to their destruction" (Isaiah 10:24-25). Furthermore, God had assured his people, "Babylon, the jewel of kingdoms...will be overthrown by God like Sodom and Gomorrah" (Isaiah 13:19). God had allowed these two powerful nations to be used as his instruments in punishing the wickedness of his people. However, although they were used by God, the Lord did not approve of their cruelty and arrogance. Once they had served their purpose, they were also destroyed for their own wickedness. Much of what we know today about Assyria and Babylon has had to be pieced together by archaeologists because the cities of Ashur, Nineveh, and Babylon were all destroyed. God's judgment was fulfilled on them as well.

LESSON 23: THE EXILE IN BABYLON

Israel's time of exile in Babylon was a very important period in the history of God's people. The exile brought about changes in the Israelites' language and in their ways of worshiping God. It was the time of Daniel and Ezekiel. We will learn more about them in this lesson.

At the end of this lesson you should be able to:

* *Tell which of Israel's prophets wrote a letter to Babylon;*
* *Describe one of Daniel's prophecies about Nebuchadnezzar and one about Nabonidus;*
* *Name the new language the people of Israel learned in Babylon;*
* *Describe three ways the exiled Jews worshiped God while in Babylon; and*
* *Name the prophet who lived along the Kebar River and describe his main prophecy.*

A. LIFE DURING THE EXILE

During the last days of the Kingdom of Judah, the prophet Jeremiah was preaching God's word of warning to God's people. After the first exiles had gone to Babylon, God told Jeremiah to write a letter to the exiles and give them these instructions:

> Build houses and settle down; plant gardens and eat what they produce. Marry and have sons and daughters, find wives for your sons and give your daughters in marriage. Increase in number there and do not decrease. Also, seek the peace and prosperity of the city to which I have carried you into exile.... When 70 years are completed for Babylon, I will come to you and fulfill my gracious promise to bring you back to this place. (Jeremiah 28:5-7, 10)

With these words, God made it clear to the people that their captivity was not going to end after a few weeks. It would last 70 years. Therefore, God wanted them to start a new life in Babylon and live as normally as possible. Yet, the Lord assured them that after the 70 years were finished, he would bring them back to the land of Israel. The exiles obeyed God's

Word. They built houses, raised families, and pursued their occupations in Babylon. Some were so successful that they did not wish to return to Israel after the 70-year period. Many others, however, though they prospered outwardly, still yearned to return to their native land.

B. DANIEL IN BABYLON

In 597 B.C., ten years before the final destruction of Jerusalem, Nebuchadnezzar of Babylon defeated Jerusalem and made it subject to him. At that time, he carried off many of the precious objects in the temple. He also commanded that some of the leading citizens, like King Jehoiakim, be taken as captives to Babylon. He especially asked his officers to bring back healthy and clever young men who could be trained as court officials (Daniel 1:1-4).

One of the young men who was taken was Daniel. Three others were given the names Shadrach, Meshach, and Abednego. They were taught to read and write the cuneiform language of Babylon (Daniel 1:4). For three years, they studied all the wisdom of the Babylonians so that they could serve as advisors to the king.

Daniel continued as an advisor to the king throughout most of the Babylonian captivity. God gave him the ability to interpret King Nebuchadnezzar's dream and to predict the king's terrible sickness (Daniel 2-4). Nebuchadnezzar died in 562 B.C., about 25 years after the destruction of Jerusalem. The book of Daniel tells us nothing about the next two kings, Amel Marduk (ruled from 562-560) and Neriglissar (ruled from 560-556). King Nabonidus came next. He officially ruled from 556-539, but he spent most of his reign in retirement at the desert oasis of Tema. He appointed his son Belshazzar as his co-ruler in 553, and Belshazzar did the ruling while Nabonidus was enjoying other pursuits. Daniel chapter 5 tells us about Belshazzar's wicked feast and how God's writing on the wall foretold the end of his kingdom. The Persians then captured Babylon, and Daniel continued to faithfully serve the new Persian king in Babylon (Daniel 6).

Daniel spoke his last recorded prophecy in the third year of Cyrus's reign (537 B.C.). We have no record of Daniel ever returning to Jerusalem. Perhaps he died before he was able to, or perhaps he was too old, or God had instructed him to remain in order to bring God's messages to the rulers of the Persian Empire. Daniel's story gives us an example of how his true followers accepted the fact of their exile, still lived a God-pleasing life, and became important instruments of God's blessings to others, all the while remaining faithful to the Lord.

C. THE WISDOM AND GLORY OF BABYLON

The city of Babylon was already over 1000 years old when King Nebuchadnezzar began his reign, but he raised the city to new heights. The city was built on both shores of the Euphrates River and was surrounded by two protective walls—one 22 feet (6.5 meters) thick, the other 25 feet (7.5 meters) thick. Three hundred sixty watchtowers were placed along the wall, each of them 90 feet (27.5 meters) high.

Three large building complexes inside the city were among the most famous in Mesopotamia. The first was the religious center for Babylon's chief god Marduk. The temple was at the top of a 300-foot (90-meter) high stepped tower called a ziggurat. There was also a vast royal palace for the king and his court (Daniel 4:29). The third structure was known as the Hanging Gardens of Babylon. These gardens were numbered among the seven wonders of the ancient world. The ancient descriptions say that the gardens were built on several levels with thousands of trees and plants which hung gracefully down its sides. These structures proved the immense wealth and splendor of the Neo-Babylonian Empire.

Babylon was known for much more than just spectacular buildings; it was also famous for its knowledge and wisdom. The Babylonians took great pride in collecting and preserving old stories from the Akkadians and other earlier civilizations. They published new editions of the stories about Gilgamesh (see Lesson 7) and added to the books written about the creation of the world and the great flood. They made new advancements in the fields of mathematics. They based their mathematical system on the number 60, a custom we still follow in dividing hours into minutes. They also had skilled astronomers who charted the positions of stars and planets and studied eclipses. Their medical experts performed surgery and studied human anatomy. Other scholars made long lists of plants, animals, and stones, while still others were able to purify gold and make glass.

These achievements clearly show why the Babylonians became famous for their knowledge. Over time, the word Chaldean (another name for the Babylonians) came to mean "wise man" or "magician." This was the knowledge that Daniel and his three friends learned, making them also *magi* or "wise men." They in return shared with their Babylonian masters the special truths of God's Word that they brought with them from Israel. It is quite possible that Daniel left copies of his own writings, as well as those of other portions of the Old Testament, in the libraries of Babylon. More than 500 years later, St. Matthew used the word *magi* to describe the men who came from the east following the new star to Bethlehem to worship the newborn baby Jesus (2:1,7). They may well have come from Babylon.

D. THE PROPHET EZEKIEL

Not all the exiles from Israel were residents of the city of Babylon. Some lived on the broad plains near the Tigris and Euphrates. Others settled on an old tell along the Kebar River, a canal which helped control the Euphrates during the rainy season. There they built a city they called Tel Aviv (Ezekiel 3:15).

One of the exiles living in this region was Ezekiel. He had been a priest in Jerusalem before he was carried into captivity in one of the early deportations. The Lord came to Ezekiel during his captivity and made him a prophet to his people in Babylon. At first, Ezekiel prophesied the final destruction of Jerusalem if the people did not repent. After the city was destroyed in 586 B.C., Ezekiel continued to bring God's messages to the people. He assured them of God's deliverance from captivity and promised that a new Jerusalem would be built someday. Ezekiel also prophesied that the temple to the true God would one day be rebuilt in Jerusalem. He also spoke about the heavenly Jerusalem which awaited all true believers with the coming of the Messiah.

E. CHANGES IN LIFE AND WORSHIP

Seventy years of living in exile brought numerous changes to the daily life and religious life of God's people. As residents of a foreign country, the Jews had difficulty practicing all the customs which made them different from other peoples. Little by little, they adapted to their new surroundings both socially and religiously. One change for God's people was being given a new name. Since they had come from the Kingdom of Judah, they became known as Jews.

An even more important change took place in their language. For hundreds of years, ever since the time of Abraham, the Hebrew language had remained basically the same. In the period of the Divided Kingdom, however, the people of Israel were exposed to the predominant language of the peoples around them—Aramaic. Aramaic, a language quite similar to Hebrew, was spoken originally by the Arameans around the city of Damascus. As their population grew, and especially as Damascus became a key center of trade in the Near East, their language was adopted by more and more people. Aramaic soon became the language not only for trade and commerce, but also for government business. Already by the time of Hezekiah (800 B.C.), Hebrew officials were conducting many of their affairs in Aramaic (2 Kings 18:26).

Under the domination of Assyria and Babylonia, Aramaic became the common language for all the nations around Israel. During the exile, as Jews lived and worked among foreign peoples on a daily basis, more and more of them learned Aramaic. Gradually it became the everyday language of many Jews. Though Hebrew was still spoken in many homes, among the educated, and in religious services, Aramaic became the more common language. Three sections of the Old Testament—half of Daniel (2:4-7:28) and two parts of Ezra (4:8-6:18, 7:12-26)—were written in Aramaic.

With the destruction of the temple, worship life for Israel experienced changes also. Unable to perform the rituals that had been such a vital part of their lives, God's people developed new means of sustaining their faith. Private prayer took on an expanded role for the Jewish exiles. Daniel was not the only Jew who prayed at regular times throughout the day (Daniel 6:1-14). Another means of maintaining their spiritual life was to read and study the Scriptures. Because of the limited number of copies, and the low level of literacy, this was usually done under the direction of a priest or a prophet. Ezekiel encouraged the Jews to gather locally for worship on the Sabbath (20:20), and these gatherings later became known as synagogues (the Greek word for "gathering"). When the Hebrew Scriptures were read in synagogue services, the priest might give an explanation in Aramaic to help the people better understand what was being said. The synagogue served as a place for worship, fellowship, and religious education. Synagogues would remain an integral part of Jewish worship life even after the temple was rebuilt in Jerusalem. By New Testament times, special synagogue buildings started being constructed as worship facilities. Today, worship at the synagogue is still the center of Jewish religious life.

The 70 years in Babylon were a time for repentance and a time of change. Living in a foreign country, God's people experienced changes in language and culture. They were exposed to new ideas and different religious views, and they could not worship as they always had at the temple. Those changes, however, did not mean that Israel had to give up the worship of the true God. Daniel and Ezekiel are examples of believers who faithfully continued to worship God even as they faced challenges in their worship life.

LESSON 24:
THE RETURN FROM EXILE

After the fall of Babylon in 539 B.C., the new Persian kings allowed the Israelites to return to Jerusalem. Groups of Jews did return at various times throughout the coming century. This was the time of the great leaders Zerubbabel, Ezra, and Nehemiah. The prophets Haggai, Zechariah, and Malachi, and the heroic Queen Esther also lived during this period. In this lesson, we will learn about the Israelites' return from Babylon and their new life in Israel under Persian rule.

At the end of this lesson you should be able to:

* *Give two names for the empire that ended the rule of Babylon;*
* *Name the king and the empire that allowed the Jews to begin their return;*
* *Name Israel's chief opponent when they returned to Jerusalem;*
* *Name the date, the leader, and the chief result of each of the three groups returning to Israel;*
* *Name the prophet who was associated with each return; and*
* *Name the Jewish woman who became a Persian queen.*

A. THE PERSIAN EMPIRE

To the east and somewhat to the north of Babylon (in the modern-day country of Iran) lived the Medes and the Persians. Around 550 B.C., the young Persian ruler Cyrus united these two peoples and set out to expand his empire by overthrowing the Babylonians. After conquering much of Asia Minor, he was ready to attack Babylon. In 539 he won a decisive victory over the Babylonian army, and soon after, the city of Babylon surrendered without a fight. Cyrus and his successors continued their conquests until Persia controlled the entire Near East from the borders of Egypt to the Persian Gulf. The Bible also mentions "Darius the Mede" as an early ruler over Persia (Daniel 5:31). There is some uncertainty about who this Darius was and how he fits into the chronology of this period. Students should consult a good study Bible or Bible reference book to explore the different explanations that have been offered.

The empire of Cyrus and his successors was the first of several Persian empires. It is referred to as the Achaemenid Empire to distinguish it from the others. The Persian Empire was even larger than the empires of Assyria and Babylon. It ruled over so many lands and so many smaller kings that the Persian ruler took the name "King of Kings" (Ezra 7:12). About 330 B.C., the Greek general Alexander the Great conquered the Persians and brought their empire to an end. Chart 10 lists the early Persian kings and the dates that each ruled. These are the rulers that most influenced the history of the Israelites.

Since the Persian kings had such a vast area to rule, they tried to win the goodwill of their subjects. This enabled them to rule more effectively and efficiently. One of the ways of gaining favor with the people was to allow all the foreign gods which had been captured and taken away by the Assyrians and the Babylonians to be returned to their original temples. They also permitted many of the exiled peoples to return to their homelands. In this way, God used the Persian kings to fulfill his promise to the Israelites that they would return to the land of Israel after their 70 years of captivity.

CYRUS II (the Great)	559-530 B.C.
CAMBYSES	530-522 B.C.
DARIUS I (the Great)	521-486 B.C.
XERXES (Ahasuerus)	486-465 B.C.
ARTAXERXES I	465-424 B.C.

Chart 10. Important Persian Kings of the Achaemenid Empire

B. ESTHER, THE JEWISH QUEEN OF THE PERSIANS

The Bible devotes a whole book to the story of Esther, an orphaned Jewish girl, who rose from obscurity and became the queen of Persia. God used Esther to play a key role in the preservation of Israel. Esther lived during the reign of the King Xerxes I, about 475 B.C. The events of the book take place at the Persian capital of Susa, and its main characters are Esther, her uncle Mordecai, King Xerxes, and one of his officials, Haman.

As the book opens, King Xerxes is searching for a new queen and chooses Esther (Esther 1-2). Later, Esther's uncle Mordecai makes her aware of an evil plot by the jealous Haman to destroy all the Jews within the Persian Empire. Esther bravely comes before Xerxes, reveals Haman's intentions, and explains the peril for herself and her people. Haman is hanged on the very gallows which he had built to kill Mordecai, and Mordecai takes Haman's place as a royal advisor (Esther 2-8). The decree to kill the Jews, however, cannot be reversed, because a "law of the Medes and Persians" could not be changed (this expression is still sometimes used today of a decision that cannot be changed). Xerxes, however, issues a new decree which allows the Jews to defend themselves, and thus the Jews are saved. The Jews remember this event each year on the festival of Purim (Esther 9).

C. THE FIRST RETURN: REBUILDING THE TEMPLE

In the first year of Cyrus king of Persia, in order to fulfill the word of the LORD spoken by Jeremiah, the LORD moved the heart of Cyrus king of Persia to make a proclamation throughout his realm and also to put it into writing:

> This is what Cyrus king of Persia says: "The LORD the God of heaven, has given me all the kingdoms of the earth and he has appointed me to build a temple for him at Jerusalem in Judah. Any of his people among you may go up to Jerusalem in Judah and build the temple of the LORD, the God of Israel, the God who is in Jerusalem, and may their God be with them. And in any locality where survivors may now be living, the people are to provide them with silver and gold, with goods and livestock, and with freewill offerings for the temple of God in Jerusalem." (Ezra 1:1-4)

With this announcement by Cyrus in 539 B.C. the Jewish captives living in Babylon were free to return to their homeland. About 50,000 Jews in Babylon returned to Palestine under the leadership of Zerubbabel, a grandson of King Jehoiachin of Judah. When they arrived, they immediately built an altar. Joshua, the grandson of the last high priest, took charge of the offerings and began conducting daily sacrifices, and the Jews joyfully celebrated the Festival of Tabernacles as they had in the past (Ezra 3:1-6).

The settlers then set about the larger task of rebuilding the temple. The neighboring Samaritans (see Lesson 22, part C) offered to help, but the Jews refused because they did not want the help of false worshipers. As a result, the angry Samaritans tried to hinder the work of the Jews (Ezra 4-6). However, with God's help, and through the encouragement of the prophets Haggai and Zechariah and the leadership of faithful Zerubbabel, they finally completed the difficult task.

On March 12, 516 B.C., the new temple was dedicated and God's people once again were able to worship the Lord at his temple in Jerusalem. Many scholars refer to the time period of Israel's history that followed, up until the destruction of Jerusalem in A.D. 70, as the Second Temple Period (the first temple was the one built by Solomon). Because of their limited resources, the returning Jews built the second temple much smaller and less ornate than the previous one. Later, however, King Herod made it larger and more grandiose than even Solomon's temple.

The returned exiles were still a small group in 515 B.C. They controlled only Jerusalem and the northern half of Judah. They were surrounded by hostile enemies—Philistines, Edomites, Ammonites, and Moabites. The Samaritans in the northern half of Israel were especially hostile. Although the Jews and these other nations were still part of the Persian Empire, the Persians did little to maintain law and order. They only appeared regularly to collect taxes. Despite their adversities and comparative poverty and weakness, the Jews had made a beginning of rebuilding their nation.

D. THE SECOND RETURN: EZRA

About 30 years after the dedication of the temple, Xerxes came to the throne of Persia and Esther became queen. Scripture tells us almost nothing else about what happened in Israel during the 50 years following the first return. The Jews who had returned were gradually becoming more prosperous, but unfortunately, the worship of the true God was becoming entangled with the sinful practices of the surrounding peoples. Many of the Jewish men had married heathen women and were succumbing to the influence of their worship of false gods. Israel was reverting to the lifestyle which had resulted in their banishment from the land a century earlier.

This sinful situation is what Ezra encountered when, by permission of King Artaxerxes I, he led a second group of about 1800 exiles back to Jerusalem in 457 B.C. Ezra was a "teacher well versed in the Law of Moses" (Ezra 7:6). His contribution to the rebuilding of Israel was spiritual as well as physical. Ezra had been authorized by Artaxerxes to appoint rulers and to oversee the teaching of the Law within Israel. He confronted the problem of Jews intermarrying with unbelievers and led the people to acknowledge their sin and end their marriages with their foreign wives (Ezra 10). Ezra's efforts resulted in a spiritual reform, and Israel once again rededicated itself to the true God.

E. NEHEMIAH AND THE THIRD RETURN

About 10 years after Ezra's return, a third group of exiles also made the journey back to Israel. This group was led by Nehemiah, a high official (the cupbearer, Nehemiah 2:1) of Artaxerxes. He was concerned about the opposition his countrymen were receiving from their enemies. He also heard that the Jews were unable to protect themselves because the walls of Jerusalem had not been rebuilt. So, Nehemiah requested a leave of absence from Artaxerxes's service so that he could oversee that important project. Carrying official orders from the king, Nehemiah went to Jerusalem and organized the rebuilding of the walls (Nehemiah 2). Judah also became a separate province from Samaria, and Nehemiah was appointed as its governor.

Nehemiah made his journey to Jerusalem in 444 B.C. When he arrived, he secretly examined the city's condition during the night. Having made his own observations, Nehemiah called on the people to unite in the task of rebuilding the wall. The people responded favorably, and they started work almost immediately and progressed quickly. Once again, however, the Samaritans attempted to stop their work. The governor of Samaria, Sanballat, led the opposition. He first tried mockery, then attempted sneak attacks, and finally tried to capture and discredit Nehemiah. However, Nehemiah and the people responded to each of these challenges. With weapons in hand, they continued to repair and reconstruct the walls, and they completed their task two months later (Nehemiah 4).

For the next two years, Nehemiah continued to strengthen and reinforce the walls of Jerusalem. During this time, Ezra was also busy leading a spiritual revival. He read the Law of Moses to all of the people gathered in the city of Jerusalem and then carefully explained it in Aramaic so that all the people could understand (Nehemiah 8:1-10). The result was another spiritual renewal among the people. The Jews then held a great celebration to commemorate the completion of the walls.

When Nehemiah's leave of absence from the king came to an end, he returned to Persia. When he visited Jerusalem again several years later, Nehemiah found that the religious life of the people was once again deteriorating. They were ignoring Sabbath laws and again intermarrying with foreign unbelievers. So Nehemiah attempted to restore true worship to the true God in Israel once again (Nehemiah 13). It was during Nehemiah's absence or second visit to Jerusalem that the last Old Testament prophet, Malachi, was also proclaiming God's Word to the people. He continued to preach against religious indifference and the corruption of Israel's worship practices. Malachi was the last of the prophets included in the Old Testament, the inspired record that God gave us about his plan of salvation.

LESSON 25: THE GREEKS AND ISRAEL

In Lesson 3 we learned about the nine major periods of Bible history. Up to this point we have studied seven of those periods. In this lesson, we begin our study of the eighth period, the intertestamental period. This period lasted about 400 years. During those years, the Greeks played an important role in Israel's history. This lesson gives us details about developments in Greece at that time and explains who Alexander the Great was and how he influenced Israel's history.

At the end of this lesson you should be able to:

* *Identify the names and give the dates of the first four rulers of Israel during the intertestamental period;*
* *Describe the major geographical features of Greece;*
* *Recount the exploits of Alexander the Great and the areas he conquered;*
* *Describe most lasting result of Alexander's conquests and its effects upon Israel;*
* *Give the name of the Greek family that ruled Egypt, tell how long they ruled Israel, and briefly describe their rule; and*
* *Give the name of the Greek family that ruled Syria, tell why the Jews hated them, and describe the result of their rule over Israel.*

A. THE INTERTESTAMENTAL PERIOD (400 – 0 B.C.)

After the time of Malachi, God preserved no other written messages for his people until New Testament times. Thus, the Bible provides no record of what transpired in this period between the Old and New Testaments. Important events, however, did occur over those 400 years. Not only in Israel, but also throughout much of the Mediterranean world, significant changes were taking place. Persia continued to rule Israel and the Near East for another 100 years until about 330 B.C. The people of Israel were prospering economically, and there were Jews across the Persian Empire who became wealthy businessmen as trade and commerce flourished. The large Persian Empire, however, was beginning to show signs of weakness that eventually resulted in its collapse. The time was ripe for a new conqueror to create a new empire.

B. THE COUNTRY AND THE PEOPLE OF GREECE

Greece is a European peninsula surrounded by water on three sides. To the west, it is separated from Italy by the Adriatic and Ionian seas. On the south, it is separated from Africa by the much larger Mediterranean Sea. And to the east, it is separated from Asia Minor (modern-day Turkey) by the Aegean Sea. Only in the north is it connected to land, and it forms one of the southern borders of Europe. Greece is also surrounded by a large number of islands where Greek culture also flourished. In fact, long before the intertestamental period, the Greeks had established colonies in Asia Minor, along the shores of the Black Sea, and in Italy as well.

On Crete, the large Mediterranean island south of Greece, the first great Greek civilization emerged. This civilization had flourished on Crete already over a millennium before the time of Abraham (3500 B.C.) and lasted until the time of the Israelite judges (1400 B.C.). This is known as the Minoan civilization.

During the latter stages of the Minoan civilization, new Indo-European peoples from the north began to move into Greece. These new migrants mixed with the existing people and formed a new culture which dominated the entire land of Greece by the period of the Judges (1600-1100 B.C.). This is called the Mycenaean period of Greek history. By that time these people were also living in many areas of western Asia Minor and the islands of the Aegean Sea. The Greeks at this time were not a united people. Rather, Greece was made up of many small city-states which often fought with each other. Each of these city-states had their own form of government and their own army, yet they shared many of the same cultural elements. Each area also spoke a Greek dialect which differed somewhat from that of its neighbors.

As the Greeks continued to expand, they came into conflict with the empire of Persia. The two greatest of Greece's city-states, Athens and Sparta, joined forces and led the Greeks in defeating Xerxes's mighty army and navy shortly after 500 B.C. In the years that followed, however, Athens and Sparta spent much of their time and resources waging war against each other. However, this period is still called the "Golden Age of Greece" (roughly 500-400 B.C.) because numerous famous historians, statesmen, artists, architects, scientists, doctors, playwrights, and poets lived and flourished during this time. The three most famous Greek philosophers—Socrates, Plato, and Aristotle—also lived in this century and the one that followed. Modern ideas of democracy trace their roots to the Greeks of this period.

C. ALEXANDER THE GREAT (336-323 B.C.)

While the Greeks were fighting among themselves, their cousins to the north, the Macedonians, became a powerful nation. They conquered Greece in about 350 B.C. In 336, Alexander, a young man of only 20, became the king of Macedonia. Because of his remarkable accomplishments, we know him today as Alexander the Great. Once he became king, he soon set out with his army to wage war against Greece's greatest enemy—Persia. His armies first drove the Persians from Asia Minor, Syria and Palestine (without bothering much with the Jews). After expelling the Persians from Egypt in 332, he was proclaimed pharaoh and founded the great city of Alexandria in the Nile delta.

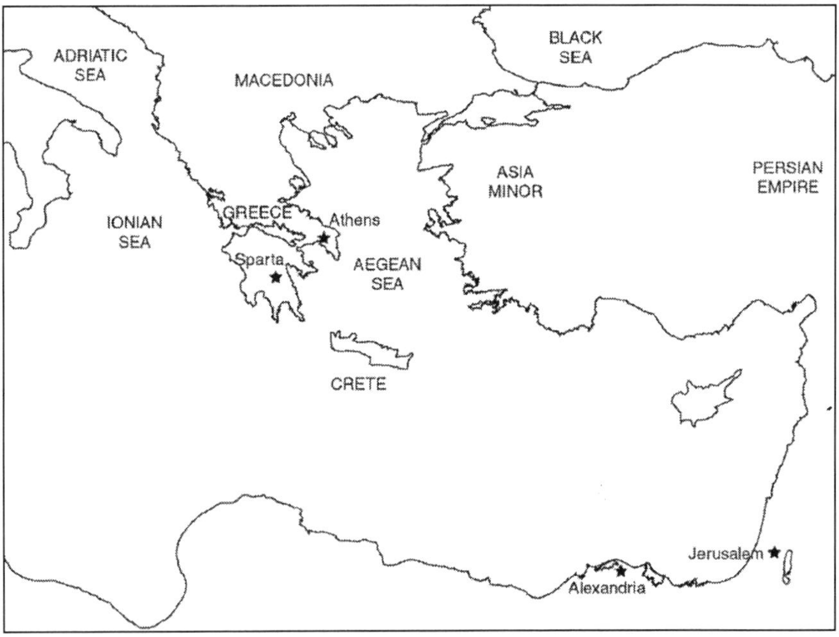

Map 21. Greece and Its Neighbors

He then turned back eastwards and attacked the Persian homelands in Mesopotamia. By 330 B.C., the Persian Empire had fallen, and much of its territory was under Alexander's control. Then, Alexander pushed his army to further conquests in Afghanistan, Pakistan, and even India. When his soldiers refused to travel further, they returned to Mesopotamia. Shortly after arriving back at Babylon, Alexander caught a fever and died (323 B.C.). At the time of his death, Alexander's empire was one of the greatest that the world has ever seen.

Wherever Alexander traveled, he left behind Greeks and Macedonians—rulers, soldiers, and traders. In this way, Greek culture and ideas and the Greek language also spread. The culture they spread was called Hellenism (the Greeks called their country "Hellas"), and the international type of Greek that was spoken in these areas is called Hellenistic or Koine Greek (common Greek). Later the New Testament books were written in this language. Although Alexander's empire began with military conquest, the legacy of Greek customs and language had more far-reaching and long-lasting effects. The similar culture and common language were also highly beneficial to the spread of the gospel following the death of our Savior.

D. ISRAEL UNDER THE PTOLEMIES (312-198 B.C.)

Upon Alexander's death, his empire almost immediately disintegrated. For more than a century, his most important generals and their successors sought to expand their own areas of rule. Ptolemy and his successors (the Ptolemies) ruled Egypt; Syria was ruled by Seleucus and his heirs (the Seleucids); and Antigonus and his family (the Antigonids) controlled Asia Minor.

With Alexander's defeat of the Persian Empire, Israel briefly came under the direct control of the Macedonians. After Alexander's death, the Ptolemies of Egypt and the Seleucids of Syria fought for control of Israel. After several indecisive battles, the Ptolemies gained control in 312 B.C. For the next century, Israel remained under the domination of this Greek dynasty which ruled from Egypt.

Under the Ptolemies, the Jews lived quite peaceably. The Ptolemies allowed them to govern most of their own affairs and to continue to worship and live as they had done before. The Ptolemies also encouraged Jews to live in various Egyptian cities and granted them special privileges for living there. The Ptolemies' capital city of Alexandria became one of the greatest cities of the ancient world. Its 400-foot (120-meter) high lighthouse was one of the seven wonders of the ancient world. Its famous library and museum attracted many scholars and made Alexandria a center of learning. During this period, the Hebrew Old Testament was translated into Greek at Alexandria; this translation is known as the Septuagint.

E. ISRAEL UNDER THE SELEUCIDS (198-165 B.C.)

In 198 B.C., however, a power shift took place. The Seleucid army of Antiochus III defeated the army of the Ptolemies near the source of the Jordan River. For the next three decades, the Jews were under the control of these Greek rulers from Syria. At first, the Jews were happy with this change of government. They received favors and privileges from their new rulers. Unfortunately, this kindness only lasted for a short period of time.

At this same time, the Seleucids were allied with Carthage, a city on the North African coast, against the rising empire of the Romans. Hannibal, the famous general of Carthage, tried to invade Rome with support from the Seleucids. When Hannibal was defeated in 202 B.C., Rome sought revenge on the Seleucids for their support of their enemies. As a result of Carthage's defeat, the Seleucids too were forced to pay huge amounts of tribute to Rome. In order to raise the money, the Seleucids increased their already heavy taxes on the peoples they ruled—including Israel. They even took gold from the temples of a number of these countries.

The Jews quickly grew to hate these new rulers and were furious when they robbed the temple in Jerusalem of some of its valuables. But the Seleucids just continued to increase their persecution of the Jews—first by increasing their taxes and then by forcing them to adopt Greek ways and culture. The Seleucids built Greek theaters, temples, race-courses, gymnasiums, and public baths in Israel's largest cities. The position of high priest was no longer given to one of Aaron's descendants, but was instead sold to the highest bidder. Such disregard for Jewish religious customs resulted in numerous uprisings. During one rebellion alone, over 40,000 Jews were killed. Finally, in 167 B.C., the Seleucid king Antiochus Epiphanes went too far. He ordered all Jews to abandon their native customs and follow Greek and Syrian practices. To show his contempt for his Jewish subjects and their religion, Antiochus sacrificed a pig on the altar of the temple at Jerusalem.

For every faithful Jew, this pagan sacrifice was a terrible abomination. Confronted with demands that were clearly contrary to God's Law, many Jews refused to obey Antiochus's orders and continued to worship according to Moses's instructions. Many Jews were tortured and killed for practicing their faith. Some of the description of persecution and perseverance of believers in Hebrews 11:35-38 seems to be recounting events of this period. These were indeed trying times for God's people. The Seleucid oppression led to an increasingly fierce nationalism among the Jews, and finally to a major revolt—that of the Maccabees.

LESSON 26:
THE MACCABEES AND THE APOCRYPHA

During the second half of the intertestamental period, the Jews regained their independence to a great extent. For almost a hundred years, the Jewish family known as the Maccabees or Hasmoneans ruled them. In this lesson we will learn how this change in circumstances came about. We will also take a brief look at the books which we call the Apocrypha, since most of them were written during this period.

At the end of this lesson you should be able to:

* *Give the date for the Maccabean revolt, and name the new Jewish festival started at this time;*

* *Give the dates for Maccabean rule, and name the two main territories conquered and the main southern people conquered during this period;*

* *Name the two main religious parties to which the Maccabean rulers belonged, and tell how these parties' attitudes towards Jewish culture differed;*

* *Tell how and by whom Maccabean rule ended;*

* *Describe the number of books, and the main literary types in the Old Testament Apocrypha;*

* *Explain the Apocrypha's relationship to the Old Testament.*

A. THE MACCABEES

At the end of our last lesson, we related that the Seleucids had tried to force the Jews to be Hellenized – insisting that they adopt Greek culture and religion. In 167 B.C., King Antiochus IV even had a pig sacrificed in the temple at Jerusalem in order to make it unclean for Jewish sacrifices. The officers of King Antiochus also went into many other cities of Israel and forced the Jewish leaders to deny their faith and sacrifice to heathen gods. In the city of Modin, an old priest named Mattathias refused to follow these orders. When another Jewish leader did sacrifice to a heathen god as the Syrians had ordered, Mattathias grabbed a knife and killed him, as well as the Seleucid officer. Then Mattathias fled into the Judean hills with his sons and friends. There, they formed a band of freedom fighters. Soon, many other loyal Jews from all over Israel joined them. Mattathias died soon afterwards, but his son Judas took his place as the leader of the rebellion. Judas, who was nicknamed Maccabeus ("the one who hammers"), became a deadly foe of the Seleucids. Judas and the family members that followed him as leaders of the revolt came to be called the Maccabees; they are also referred to by their older family name as Hasmoneans.

For the next 30 years (about 165-135 B.C.), the Maccabees fought to gain freedom for Israel from the Seleucids. Slowly, they regained control of more and more territory from the Seleucids. Judas defeated the enemy in several important battles. In 164 B.C., he gained control of the city of Jerusalem. He cleansed the temple and rededicated it to the worship of the true God. The Jews have commemorated this event ever since in their festival called Hanukkah (or Dedication). Hanukkah quickly became one of the great Jewish festivals (John 10:22).

Judas, however, only succeeded in freeing portions of Jerusalem and other small regions of Judea from Seleucid control. Once that had been accomplished, many of Judas's supporters deserted him. Since they could now worship freely again at the temple, they did not wish to fight any longer. Judas was also killed shortly afterwards in 160 B.C. Command of the rebels then passed to Judas's brother, Jonathan.

Though Judas had gained freedom for the temple in Jerusalem, the Seleucids still controlled other areas of the city. During a period of Seleucid political weakness, while several men were claiming to be the Seleucid ruler, Jonathan was appointed high priest by one of the Seleucid claimants. Through clever political maneuvering, Jonathan was able to increase his own power in Israel and gradually reduce Seleucid influence. Jonathan eventually gained control of all Jerusalem.

After Jonathan was treacherously killed in 142 B.C., his brother Simon took over leadership. Through a treaty with the new Seleucid king, Simon was able to free Israel from all Seleucid taxes and control. From about 140 B.C., the Jews were treated as an independent state allied with the Seleucid Empire. Thus, Israel finally achieved control of its own destiny, and this is usually given as the start of the Hasmonean Kingdom. Simon then attempted to rebuild his country after the many years of fighting. He extended Israel's power by driving his enemies out of the western foothills and also gained control of all the land between Jerusalem and the Mediterranean Sea. Israel became more prosperous than it had been for many years. However, Simon's rule did not last long and ended with his own violent death in 135 B.C.

B. THE LATER MACCABEANS

Simon was succeeded by his son John Hyrcanus. John also became the high priest. Since the Seleucids did not allow the Jewish rulers to use the term king to describe themselves, they merely ruled under the title of high priest. John was able to extend Jewish rule to include Samaria and Galilee in the north, as well as the Negev in the south. He destroyed the Samaritans' temple on Mt. Gerazim. He also forced the Idumeans (formerly the Edomites) to be circumcised and to adopt the Jewish religion. The Idumeans were the tribe from which Herod the Great's family would come. During this period the Jews split into several religious factions (we will study this in our next lesson). John Hyrcanus died in 106 B.C.

The rule of the Jews then fell upon John Hyrcanus's two sons—Aristobulus I (ruler from 104-103 B.C.) and Alexander Jannaeus (ruler from 103-76 B.C.). Both were rulers who used murder and coercion for their own political and personal ends. The Jewish people soon grew to hate them. Also, both favored the party of the Sadducees and persecuted those who belonged to the sect of the Pharisees (see the next lesson). When Jannaeus died, his wife Alexandra (who had previously been the wife of Aristobulus I) became the ruler of the Jews. She favored the Pharisees and introduced some important reforms. She also welcomed back Jewish exiles, restored religious ceremonies to their proper observances, collected a tax of half a shekel to support the temple, and reorganized the Sanhedrin to serve as a court of justice and to oversee religious matters. Thus, she reversed the policies of her two husbands who had encouraged Hellenism. She also encouraged the revival of a truly Jewish culture and elevated spiritual priorities to their rightful place.

After Alexandra died (in 67 B.C.), her sons Aristobulus II and Hyrcanus II fought for control of the throne. To settle the matter, the brothers appealed to Rome. But the brothers did not completely follow Rome's decision, so Rome used this as an opportunity to send its army into Israel for the first time, thus ending the period of Jewish freedom (63 B.C). Rome divided the land of Israel into several sections – Judea, Samaria, Galilee, etc., and appointed a ruler over each. Rome then made the Idumean (or

Edomite) ruler Antipater, father of Herod the Great, governor of Judea, thus beginning a new phase of Israel's history.

C. THE APROCRYPHA

As we learned earlier, no inspired books were written during the intertestamental period. However, many other books of Jewish history, poetry, proverbs, and religious stories were written during this period, most of them in the Greek language that was becoming increasingly popular. Just as the Jews of Alexandria read the 39 inspired books of the Old Testament in Greek (the Septuagint), these other intertestamental religious books became popular for those who read Greek. These books later became known as the Apocrypha (which means "hidden" or "secret" works).

Soon Greek readers were no longer clearly distinguishing the Old Testament books and the Apocryphal books. Many treated them all as part of the inspired Old Testament text. To avoid confusion, the ancient Bible scholar Jerome suggested that the books of the Apocrypha be placed in a separate place between the Old and New Testaments. This advice was finally followed by Martin Luther in his German translation of the Bible in the 1500s. Since Luther's time, Protestant churches have clearly taught that the Apocryphal books are not inspired. The Roman Catholic Church and Eastern Orthodox churches, however, still accept them as equal to the books of the Old and New Testaments and often print them in their Bible translations.

Even though its books are not inspired, the Apocrypha does provide us with interesting reading. The two books of Maccabees, for example, provide important history for the intertestamental period. The Apocrypha also helps us understand the changes that were taking place in Jewish culture and religious life. Even a cursory reading of these books, however, shows they are sometimes fanciful and inaccurate. Listed below are the 14 books of the Apocrypha in the order in which they are usually found. A summary of each book's contents is also given to provide you with some idea of the content of these writings.

I EZDRAS: Ezdras is the Greek form of the name Ezra. This book retells and adds on to the story of the book of Ezra.

II EZDRAS: This book was written about 50 years after Christ's death. It contains supposed visions of the future and predicts the Jew's rejection of the Messiah and concludes with some legendary stories of Ezra's work.

TOBIT: This book is a religious story about a man named Tobias and his son Tobit. Because of their obedience to God, Tobias is healed of his blindness, and Tobit is saved from an unpleasant death.

JUDITH: This book is a story about a Jewish girl named Judith. When the Assyrians invaded Israel, Judith charmed the enemy general and killed him, thus saving her people from conquest.

ADDITIONS TO ESTHER: This book contains six additional chapters to the book of Esther. These chapters do not add any new parts to the story, but they attempt to give the story of Esther a more religious character.

WISDOM OF SOLOMON: This book claims Solomon as its author, but it was actually written centuries later. This book tries to show that wisdom is the one true path to being a faithful Jew.

ECCLESIASTICUS: This book also provides its readers with wisdom for living a godly life. It was highly valued within the early church for the insights it gave into carrying out God's Law. It is sometimes called the WISDOM OF SIRACH.

BARUCH: This book claims Jeremiah's secretary Baruch as its author. It contains a letter, a prayer of confession, and songs of comfort.

SONG OF THE THREE CHILDREN: This book is supposedly an addition to the book of Daniel, chapter 3. It gives us the prayer and song of the three men as they walked about in the fiery furnace.

SUSANNA: This book is another addition to the book of Daniel. It gives a fanciful account of how Daniel supposedly rescued a beautiful girl by proving that the men accusing her of adultery were lying.

BEL AND THE DRAGON: This book is yet another adaptation to the story of Daniel. It explains how Daniel exposed the Babylonian gods and their priest as fakes.

THE PRAYER OF MANASSEH: This book supposedly contains the prayer which the wicked Judean king Manasseh made after he was imprisoned by the Assyrians and repented of his evil ways.

I MACCABEES: This book gives us the historical background for the Maccabean revolt against the Seleucids and describes the events of the years 175-134 B.C. fairly accurately.

II MACCABEES: This book covers the same period of history as I Maccabees. However, it varies somewhat from 1 Maccabees and was written much more from the point of view of the Pharisees. It is less accurate than the first account.

In addition to these 14 books of the Apocrypha, many other Jewish religious books were written during the intertestamental period and well into the New Testament era. Those other books are usually called *pseudepigrapha* (which means "false writings"). Many of those works claim to be the writings of Bible characters, but their content and the fact that many of them first appeared many years after the death of their supposed authors, clearly show that they are forgeries.

For about a hundred years, during the rule of the Maccabees, or Hasmoneans, Israel was once again a free nation. The 100-year reign of the Maccabees began with Israelites uniting to win their freedom, but it soon degenerated into a period of internal religious disputes and political factions. The books of the Apocrypha give us a glimpse into the history of this period this time. The apocryphal books also show us how Jewish culture was changing in the final centuries before the coming of the Messiah. We will study some of the most important religious changes in the next lesson.

LESSON 27:
PHARISEES AND SADDUCEES

In the New Testament, we read again and again about the Pharisees and the Sadducees. Yet, these religious parties never appear in any of the Old Testament books. The reason is that these religious sects were first formed during the intertestamental period. In this lesson we will briefly study some of the main religious parties which developed during this period and shortly afterwards. We will learn what each of these religious groups stood for and how they differed from the others.

At the end of this lesson you should be able to:

* *Give the names of the religious group at the time of the Maccabees and the names of the three religious groups formed soon after;*
* *Describe the most important teachings and customs of these groups and on what they based their teachings;*
* *Name the allies of the Pharisees and also those of the Sadducees;*
* *Name and describe the important religious settlement near the Dead Sea; and*
* *Give the name for the manuscripts discovered nearby and describe their importance.*

A. THE HASIDIM

We have already learned that during the Seleucid and Hasmonean periods attempts were made to force the Jews to adopt Greek culture and customs and not follow strictly all the Jewish religious beliefs and customs. During that period, a group of Jews who were faithful to their traditional culture and religion began to band together to oppose all efforts to Hellenize the Jews. This culturally and religiously conservative group became known as the Hasidim (the "separate ones").

The Hasidim were concerned with remaining true to the Law of God and the cultural traditions of Israel. They were quick to join with the Maccabees in their fight against the Seleucids. They saw the revolt as a holy war. Once the Jews had regained control of the temple, however, they refused to fight further for political freedom. When Jonathan was appointed as high priest, he again was able to win the support of the Hasidim.

The Hasidim placed a strong emphasis on the importance of the Law and insisted on following the many religious traditions that had been handed down over the years. In this way, they were the forerunners of the Pharisees. The question still arose, however, as to whether or not a person could adopt any of the Greek customs and still remain a faithful Jew. The Jewish religious leaders split over the proper answer to that question, probably during the rule of John Hyrcanus. In this way the Hasidim split into Pharisees, Sadducees, and other religious groups. While these groups had no formal membership lists, people supported the group which best represented their own political and religious beliefs.

B. THE PHARISEES

The largest group of Hasidim eventually became known as the Pharisees. This religious sect became the most numerous and most influential of the religious parties in Israel. For this reason, the Pharisees are mentioned in the New Testament more than any other group. The Pharisees' belief was centered in the Law of Moses and the many traditions that had developed around it. They obeyed all the laws about worshiping in the temple. Surprisingly, however, few priests belonged to the party of the Pharisees.

Because of the Pharisees' emphasis on the Law, however, many teachers of the Law favored the doctrines of the Pharisees. In the Old Testament, these men were called "scribes" because their main work was making copies of the Old Testament. This copying process helped them become very knowledgeable in the Scriptures. Thus, the scribes gradually came to be viewed as teachers and experts in the Law. By New Testament times, the teachers of the Law, just like the Pharisees, had become great defenders of the Law and the traditions of the elders. The "Pharisees and teachers of the Law" are frequently mentioned as presenting a united front in opposing Jesus and his teachings (Matthew 12:38).

The Pharisees also became well known for their strict outward observances of God's Laws. Since the Old Testament required every Jew to remember the Law at all times and live his whole life according to it, the Pharisees followed these instructions literally. Jesus described many of their customs in his sermon recorded in Matthew 23. The Pharisees tied pieces of

Scripture to leather straps and then tied the straps to their hands and foreheads. These straps were known as phylacteries (v. 5). In addition, the pious Pharisees were strict observers of the command to tithe (i.e., give one tenth of their income to the Lord). They tithed even the smallest things which their garden produced (vv. 23-24). They wore highly decorated prayer shawls and were often given seats of honor at banquets and in the synagogues (vv. 5-6). Because of their knowledge of the Law, they were often called "Rabbi" or teacher (v.25). They took pride in obeying all the ceremonial washings (v. 25) and gave special honor to the Old Testament prophets (v. 29). They were especially careful to observe all the rules about the Sabbath (Mark 2:23-3:6). In addition, they refused to have any contact with anyone who was not a Jew. Such people were considered religiously "unclean" and were avoided at all costs. Yet the Pharisees did try to win others to their own faith (Matthew 23:15).

In all of these ways, the Pharisees attempted to keep perfectly both God's Law and the hundreds of traditions that had been added to it. In many cases, however, they ended up keeping the exact wording of the Law while not understanding or following its true intent. Instead of seeing their own sinfulness and inability to keep the Law, they became proud of their supposed superior ability to obey the small rules (Luke 18:10 ff.). The Pharisees were also quite good at using their traditions and their own interpretations of God's Law to get around its true meaning so that the Law served their own purposes. Mark 10:1 ff., for example, shows how some Pharisees misused God's message regarding marriage and divorce.

Although at times they distorted or misread Scripture, the Pharisees often did teach the truths of biblical morality. They also correctly emphasized the Bible's teaching about the final resurrection of the body, and that rewards would be given to the righteous and punishment to the wicked. They also looked for the coming of a Messiah, but they were expecting a Messiah who would have an earthly as well as a spiritual kingdom. Their longing for a restoration of Israel's earthly glory distracted their attention from the promises about the coming Christ and his eternal kingdom.

Scholars disagree about the numbers and popularity of the Pharisees in Jesus's time, but they seem to have had supporters among all classes of society, among the teachers of the Law, and among members of the Jewish ruling council called the Sanhedrin. Nicodemus (John 3:1-21) and Saul

(Philippians 3:2-11) were two prominent Pharisees who became followers of the true Messiah, Christ.

C. THE SADDUCEES

The Sadducees were the second most influential religious sect in Israel during the New Testament period. In contrast to the Pharisees who sought to maintain Jewish religion and culture in their traditional forms, the Sadducees favored compromises to accommodate Jewish customs to the more modern Hellenistic culture. This willingness to adapt to Greek customs led to religious compromise on the part of the Sadducees as well.

The Sadducees, like the Pharisees, held the five books of Moses in high regard. For them, this was the only inspired source of teaching. The result for them, however, was a less rigid acceptance of the traditions which had been passed down by the Jewish elders. They also rejected some of the Old Testament teachings which they felt were later additions to the writing of the Bible, especially the existence of good and evil angels and the resurrection of the body after death. The Sadducees' denial of the resurrection of the body led them to oppose the Pharisees and the early Christians. They argued with Jesus on this point and tried to trap him into confessing their point of view (Matthew 22:23 ff.). Peter and John were later arrested by the supporters of the Sadducees because they were preaching that Jesus rose from the dead (Acts 4:1-2).

Whereas the Pharisees were friends with the scribes and the teachers of the Law in the local synagogues, the Sadducees allied themselves with the high priest and the other temple officials (Acts 5:17). The Sadducees generally were the more educated and upper-class citizens of Israel. During most of the New Testament period, the high priest was a Sadducee. Since the Sadducees allowed for a certain degree of Hellenization to take place among the Jews, it was easier for them to work together with foreign rulers, such as the Romans. Thus, they were able to maintain their power base and be a source of political influence. Although the sect of the Sadducees was much smaller than the sect of the Pharisees, the Sanhedrin contained an almost equal number of each. Later on, Paul used this to good advantage when he was tried before the Sanhedrin for proclaiming Christ's teachings in Jerusalem (Acts 23:6 ff.).

Both the Sadducees and Pharisees agreed that salvation was at least partially earned by good works. Therefore, as false teachers, Jesus placed them both into the same category. He warned his disciples, "Be on your guard against the yeast of the Pharisees and the Sadducees" (Matthew 16:6). Jesus was really warning them to beware of "the teaching of the Pharisees and the Sadducees" (v. 12). As a result, the Pharisees and the Sadducees, who were usually enemies, joined together to bring about Christ's death.

D. THE ESSENES, QUMRAN, AND THE DEAD SEA SCROLLS

A third religious party also existed at this time, although it is never mentioned in the Bible. This group was called the Essenes. Other historical evidence tells us that they were already in existence by 150 B.C. The Essenes were a group of Jews who believed true believers should not live in a sinful society. Therefore, they left Jewish society and formed their own isolated communities. They are comparable to the monks who formed their own religious communities in the early Christian church. The Essenes added to Scripture their own set of rules for living. Some practiced celibacy, yet their sect endured by adopting children into their groups. They preferred to live in desert areas, as far from others as possible, and at least some of the Essenes lived near the Dead Sea. They wore simple clothing, ate their meals together, and had no private possessions. Their teachings were so secretive that we still know only very little about them.

Near the middle of the twentieth century, archaeologists discovered an ancient settlement on the shores of the Dead Sea at a place now called Qumran. Some scholars think that this settlement was home to an Essene community. In nearby caves, the famous Dead Sea Scrolls were discovered. These ancient scrolls were hidden in the caves about A.D. 70. The numerous scrolls contain large portions of the Old Testament, apocryphal books, and other books which appear to have been written by Essene authors. Besides giving us some insight into the Essenes, the Dead Sea Scrolls also bear witness to the fact that the text of God's Word which we have today is a faithful preservation of what God's people had 2000 years ago.

During the intertestamental period, the Jews split into religious sects, including the Pharisees, the Sadducees, and the Essenes. This split was encouraged by the increasing pressures on Jews to conform to the Hellenistic culture that was spreading throughout the Mediterranean. In our next lesson, we will learn about two more groups—the Zealots and the

Herodians—who had different solutions to these same problems. Many other Jews probably did not consider themselves members of any religious party. They simply practiced their religion more or less rigorously in their synagogue and at the temple.

LESSON 28:
THE HERODS AND THE ROMANS

After 63 B.C., the Romans ruled Israel. Sometimes this was done through officials sent from Rome; at other times, Rome allowed local individuals, such as the half-Jewish family of Herod, to administer the affairs of the Jewish state on Rome's behalf. In this lesson, we will learn about the rule of the Romans and the Herod family in greater detail.

At the end of this lesson you should be able to:

* *Give the date for the beginning of Rome's rule in Israel and also for the rule of Herod the Great;*

* *Name the son of Herod the Great who beheaded John the Baptist and the son who was involved in Jesus's trial;*

* *Give another name for the Roman governors and identify the one who became most famous;*

* *Tell for what Herod Agrippa I and II are remembered in the New Testament;*

* *Name two Jewish revolts against Rome and when each one took place; and*

A. ROME AND ISRAEL

Founded about 750 B.C., the Italian city of Rome emerged as an important power by the 5th century B.C. The Romans first gained control of Italy and the surrounding islands. Then they began to took control of northwestern Africa and Spain. Following the death of Alexander the Great and the steady decline of the major parts of his empire, the Romans moved eastward as well and gradually gained influence over Egypt, Greece, and Asia Minor. As they expanded their control, the Romans preferred to rule through a local ruler or dynasty whenever possible. These puppet kings were allowed to rule their territory as long as they remained loyal to Rome, preserved peace, protected Roman citizens, and collected the prescribed amount of taxes for the Roman state. The Romans normally allowed countries under their rule to keep their own religion and customs as long as they peacefully submitted to Roman rule.

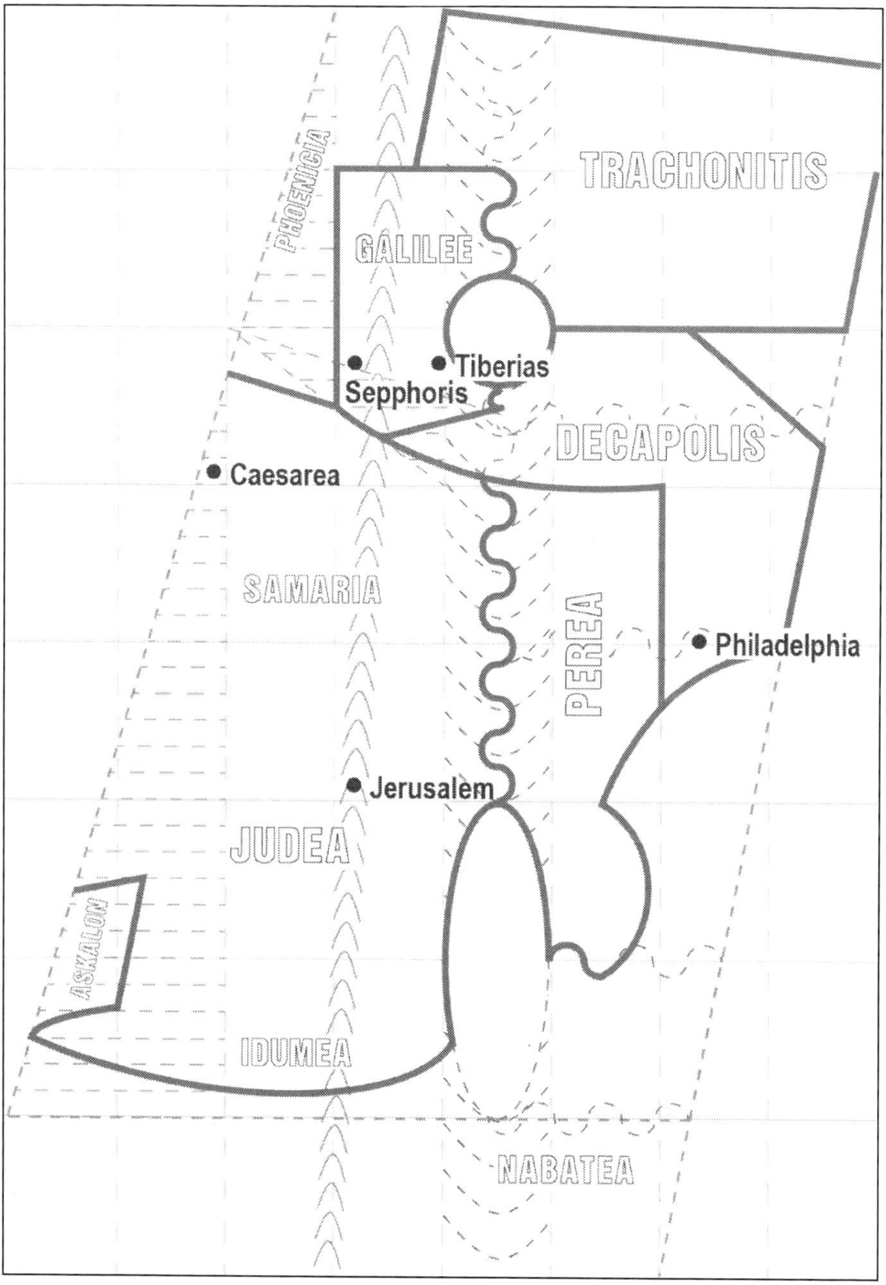

Map 22. Israel Under Roman Rule

As we learned in Lesson 26, the death of the Hasmonean ruler Alexandra in 67 B.C. plunged Israel into civil war as her two sons fought to be her successor. Since Israel had previously signed treaties of friendship with the Romans in the time of the Maccabees, some Jewish leaders appealed to Rome to settle the dispute. As Roman armies were at that time in the process of conquering Syria, Israel's neighbor to the north, they were happy to intervene, hoping to make the wider region more peaceful. When the two Jewish rivals refused to abide by Rome's decision, Rome quickly occupied Jerusalem and made Israel a part of its empire in 63 B.C.

As was their customary practice, the Romans tried to find a competent and loyal local ruler to put in charge of Israel. At first, they tried some of the remaining Hasmoneans, but none were successful. They then turned control of the area of Judea over to a man named Antipater. Antipater was the ruler of Idumea (formerly Edom), the region southeast of Israel. He became a loyal subject of Rome and ruled over Judea from 63-43 B.C. Antipater became good friends with Julius Caesar and gained Roman citizenship through this friendship. Caesar also allowed his friend Antipater to refortify Jerusalem. Although the Jews continued to worship freely at the temple in Jerusalem, the Jews viewed Antipater as a foreigner who represented Roman domination. Therefore, many Jews despised and hated him. Though he outwardly attempted to follow Jewish customs by being circumcised and following some of the Jewish religious observances, Antipater never won the support of most Jews.

Antipater had two sons whom he appointed to important positions. One was made the military leader in Judea, and the other, later known as Herod the Great, governed Galilee. Herod impressed the Romans by breaking up a band of Galilean outlaws. However, like his father, Herod remained unpopular with the Jews. In 43 B.C., Antipater died, and shortly thereafter the Parthian army captured most of Judea (Parthia was the new empire in Mesopotamia that took over after the Seleucid empire fell to Rome). Herod fled to Rome and, while there, was appointed King of Judea (40 B.C.).

B. HEROD THE GREAT (40-4 B.C.)

Even though Herod was then officially King of Judea, he still had to win back control of the region from the Parthians, which took him three years to do. In 37 B.C., he finally captured Jerusalem. He then had all the members of the Hasmonean family killed so that no one else would lay claim to his rule.

Nevertheless, Herod came to be known as Herod the Great, and he was a resourceful, efficient, and effective ruler in many ways. He is remembered as one of the Near East's great builders because he rebuilt the old cites of Samaria and Jericho and erected new cities like Sepphoris and the port city of Caesarea. He also built forts, aqueducts, palaces, amphitheaters, and a new and enlarged temple in Jerusalem. Herod also was efficient in maintaining control over his subjects. Because he gave stability to this often troublesome region, Rome allowed him to extend his territories. Eventually he ruled over most of Israel.

Although Herod might be considered a great ruler in many ways, he was extremely ruthless. He was always afraid that someone might seek to overthrow him. He killed many of his own family members, even his own wives and sons, because he suspected they were plotting against him. This was also the unstable man who had all the babies of Bethlehem killed after the magi told him that a new king had been born to the Jews (Matthew 2:1-18). Herod died in 4 B.C., a lonely, sick, and paranoid man.

C. THE SONS OF HEROD THE GREAT

Herod had arranged with the Romans that, following his death, three of his sons would divide his kingdom. His son Archelaus became king of Judea (ruling over Judea, Samaria, and Idumea). He was a cruel and incompetent king whom the Jews feared and hated (Matthew 2:22). At the people's request, Rome replaced him with a Roman governor.

A second son, Herod Antipas, became ruler over Galilee and Perea with the title "tetrarch." This is the Herod who stole his brother Philip's wife Herodias and also had John the Baptist beheaded (Mark 6:17 ff.). Jesus called Herod Antipas a fox (Luke 13:32) and warned his disciples to beware of him (Mark 8:15). He was also the man to whom Pilate sent Jesus for examination before his execution on Good Friday (Luke 23:7 ff.). The Roman emperor Caligula finally deposed and banished Herod Antipas in A.D. 39.

Herod's third son, Philip, became tetrarch of Trachonitis (see Map 22). This was not the Philip who was married to Herodias. From the little we know about Philip, he seems to have been a mild and just ruler, but he did not follow the Jewish faith at all. At times Jesus took his disciples into Philip's territory to escape the crowds (Matthew 16:13). Philip is mentioned only once in the New Testament (Luke 3:1).

D. THE ROMAN GOVERNORS

After Herod the Great's son Archelaus was deposed in A.D. 6, Rome sent out its own men to rule. Because Judea was a relatively small region, its Roman officials were subject to the governor of the larger province of Syria. The top official was called the prefect; though by the Apostle Paul's day he was called a procurator (English translations often just say "governor"). The seat of government was in the new city of Caesarea which Herod the Great had built along the coast of the Mediterranean. The prefect often traveled to Jerusalem as well, especially during the feasts, and sometimes spent the winter there.

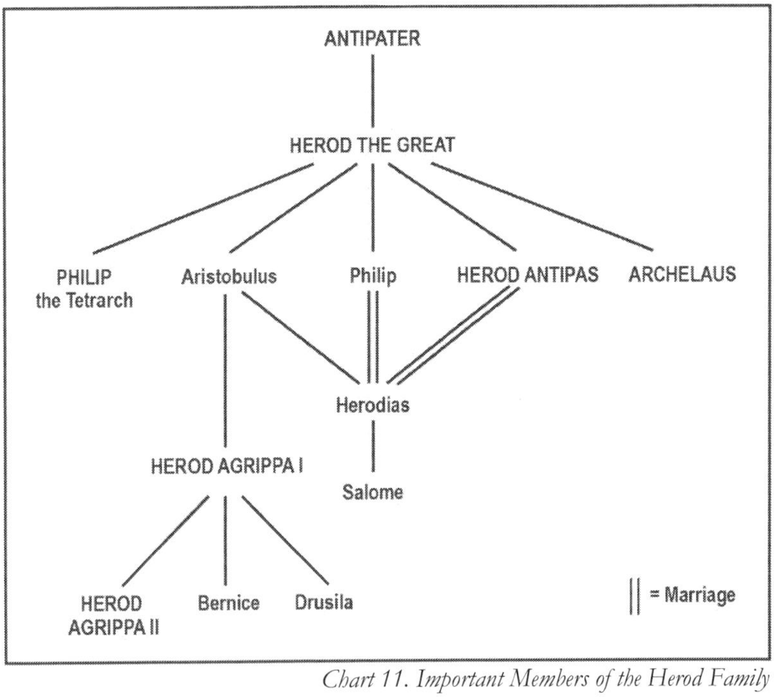

Chart 11. Important Members of the Herod Family

The prefect's main duties were to keep the peace, head up the justice system, protect Roman citizens and their interests, and make sure that the taxes were properly collected and sent on to Rome. The prefect or procurator usually allowed the high priest and the Sanhedrin to take care of most of the normal business affecting the people, but the prefect had to approve the selection of a new high priest. He also was the only one with power to impose the death penalty (John 18:31). The fourth prefect of Judea was Valerius Gratus, who ruled from A.D. 15-26. He removed the high priest Annas (John 18:13) from office and appointed his successors, including Joseph Caiphus, Annas's son-in-law (Matthew 26:57 ff.).

The fifth prefect of Judea was Pontius Pilate. He ruled from A.D. 26-36 and was the person in charge of Jesus's trial and condemnation (Matthew 27:2 ff.). The system of Roman governors lasted from A.D. 6-41. In A.D. 41, Herod Agrippa was made king of Judea, and he ruled there until his death in A.D. 44. Roman governors then ruled again until A.D. 66—the year in which the Jews rebelled against Rome. This rebellion is known as The First Jewish Revolt. By A.D. 70, Rome had crushed the revolt and completely destroyed Jerusalem along with the magnificent temple built by Herod. The Roman general Titus, who later became the Roman emperor, was responsible for the capture of Jerusalem. Titus's victory ushered in another period of strict Roman rule. This ended in The Second Jewish Revolt (A.D. 132-135). The Romans crushed the Jews just as fiercely as they had the previous time and brought an end to the nation of Israel. A large number of Jews were scattered and dispersed following these two revolts. Jews would not again rule a part of Israel until the 20th century (1948). Among the leaders of these revolts against Rome were a sect known as the Zealots. The Zealots were fanatically devoted to the idea of a free Jewish state. They stopped at nothing in their drive to rid Israel of the hated Romans. Many Zealots were willing to use terror and bloodshed to achieve their goals. It is interesting that Jesus chose Simon, a Zealot or former Zealot, as one of his disciples (Luke 6:15).

E. OTHER MEMBERS OF THE HEROD FAMILY

Two other members of Herod's family played a part in New Testament history—Herod Agrippa I and Herod Agrippa II. Herod Agrippa I, already mentioned above, was a grandson of Herod the Great. He was raised in Rome, indoctrinated into Roman ideals, and groomed for the life of a politician. In A.D. 37, when his close friend Caligula became the Roman emperor, he made Herod Agrippa I king over the territory northeast of

Israel where his late uncle, Philip the Tetrarch had ruled. When Herod Antipas was deposed in A.D. 39, Galilee was added to Agrippa's rule. When Claudius followed Caligula as emperor, he added the territories of Judea and Samaria to Herod Agrippa I's territories, making him king over virtually all of Israel. Herod Agrippa I tried hard to win the loyalty of the Jewish religious leaders. He arrested numerous Christians and even killed the apostle James, the brother of John, in order to please the Jewish religious leaders (Acts 12:1-4). As a result of his pride and arrogance, "an angel of the LORD struck him down, and he was eaten by worms and died" in A.D. 44 (Acts 12:23).

Herod Agrippa I's son Julius Marcus Agrippa, often called Herod Agrippa II, was appointed tetrarch of Trachonitis in A.D. 52. He was later given rule over part of Galilee as well. His sister Drusilla married Felix, the Roman governor of Judea, who was responsible for Paul's imprisonment (Acts 24:27). Herod Agrippa II and another sister, Bernice, went several years later to visit Festus, the new governor of Judea, and there had an interview with the Apostle Paul (Acts 25:13 ff.).

Chart 11 shows the family tree of the Herod Family. Many of these people played important roles in the New Testament. Most of them were poor rulers who were primarily concerned with their own success and prosperity, and most of them failed to gain the favor of their Jewish subjects. Some Jews, however, felt that it was better to be ruled by these descendants of Herod than by Rome directly. This group of Jewish supporters became known as the Herodians. Several times the New Testament authors link the Herodians with the Pharisees—in trying to trap Jesus (Mark 12:13 ff.) and also in plotting his death (Mark 3:6). The Pharisees saw Jesus as a threat to their religious authority. The Herodians saw Jesus as a threat to their political ideas.

LESSON 29:
MONEY IN THE BIBLE

How rich was Solomon? How much money did the poor widow throw into the collection box in the temple? How much money was needed to pay the temple tax? Money plays an important part in many Bible stories. Yet, when we read about shekels and talents, denarii and drachmas, what amounts and types of money are these? As we move from our study of the intertestamental period to the New Testament period, we will take a break from history in this lesson to discuss the different kinds of money used in the Old and New Testaments.

At the end of this lesson you should be able to:

* *Describe three steps in the development of money;*
* *Name the three main monetary units used in the Old Testament and their relation to one another;*
* *Describe the three types of money used in New Testament times;*
* *Name the most common New Testament coin and its value; and*
* *Describe the reasons money changers were found in the temple.*

A. BUYING, SELLING, AND THE VALUE OF MONEY

In the earliest civilizations, people bought and sold without the use of money. Goods were simply traded or bartered between the people involved. However, this was not always practical or equitable. People quickly found that setting some standard value on goods being exchanged was better than simply having a completely free exchange of different things. At a very early date, gold and silver already were used as a medium for trade because most people highly valued these two rare metals. Therefore, they became the standard which was used in buying and selling. At first, the metals were simply weighed to determine their value. The quality of the gold or other precious metal was also used in determining its value. By the time of Babylon's conquest of Jerusalem (about 600 B.C.), a new type of exchange was beginning. A king or ruler took a certain standard amount of gold, silver, bronze, or copper, and put his seal upon it. This seal guaranteed that the piece was the proper weight and quality. This practice spread rapidly, and the result was money as we know it.

B. MONEY IN OLD TESTAMENT TIMES

The early chapters of Genesis tell us nothing about buying and selling. Not until we come to the story of Abraham do we read anything about trade. By that time, people were already using gold and silver for commerce. The standard weight was called a shekel. The shekel remained the most common weight throughout the Old Testament period. The shekel was a unit of weight for both gold and silver. During Old Testament times a shekel seems to have weighed about 1/3 of an ounce (ten grams), yet, the weight of a shekel varied from place to place and from time to time. Thus, when Moses, the author of Genesis, recounted how Abraham paid 400 shekels of silver to buy a cave in which to bury his wife Sarah, the writer added that the 400 shekels which Abraham paid were "according to the weight current among the merchants" at that time and place (Genesis 23:15-16).

Gold was also highly valued in Abraham's time. Eliezer, Abraham's servant, gave Rebekah a gift of two gold bracelets weighing ten shekels (Genesis 24:22). A smaller unit, the beka, is also mentioned in that verse. A beka was equal to half a shekel (Exodus 38:26). The shekel could also be divided into 20 gerahs (Numbers 3:47). Or, we could say ten gerahs equal a beka, and two bekas equal a shekel. But how valuable was a shekel?

Determining the value of a shekel of gold or silver in Old Testament times is difficult, partly because values were constantly changing. Just as in modern times, inflation often affected the buying power of money. At other times, food shortages brought prices up. For instance, when the city of Samaria was being attacked by Ben-Hadad of Damascus, food prices rose rapidly within the city. The starving people were willing to pay 80 silver shekels for the head of a donkey (normally unsellable) and five shekels for 3/10 of a liter of seed pods (2 Kings 6:25). When Elisha predicted that prices would drop dramatically by the next day, the people laughed at him (7:1-2). Yet it happened as Elisha had predicted, and by the next noon seven liters of flour and 14 liters of barley each cost only one shekel (v. 16).

The price of land is mentioned several times in the Old Testament. We have already mentioned Abraham's purchase of land to bury Sarah for 400 silver shekels. David purchased a threshing floor from Araunah for 50 silver shekels (2 Samuel 24:24), although the price for the entire area and all the equipment totaled 600 gold shekels (1 Chronicles 21:25). Jeremiah bought a field for 17 silver shekels (Jeremiah 32:9). We also have information on the price of slaves. The Ishmaelites bought Joseph for 20 shekels of silver (Genesis 37:28).

The ancient monetary system also experienced a problem with cheating. People could easily weigh gold improperly and thus not give the correct amount required. Amos mentioned such cheating as common practice among the people of his day and told how the Lord would punish those who do it (Amos 8:5). By David's time, the government of Israel had established a "royal standard" (2 Samuel 14:26) by which all weight was judged. A person could check his own weights and then carry them with him in a bag on his business trips to verify that he was not being cheated by others (Proverbs 16:11). By the time of the United Kingdom, a larger weight called the mina was also in use. The mina weighed 50, or sometimes 60 shekels (Ezekiel 45:12). Thus a mina weighed about one pound (0.5 kilograms). Solomon's palace was decorated with 300 small ornamental shields, each of which was made of three minas of gold (1 Kings 10:17).

1 shekel	(10 grams; 1/3 ounce)			
1 mina	(1/2 kilo; 1.1 pounds)	= 50 or 60 shekels		
1 talent	(30 kilos; 66 pounds)	= 60 mina	= 3000 shekels	

Chart 12. Old Testament Monetary Units

The talent was an even larger weight. Sixty minas (or 3000 shekels) made up one talent. The weight of a talent was about 66 pounds (30 kilograms). When David defeated the Ammonites, he took the king's crown for himself, which weighed a full talent (2 Samuel 12:30). At the height of his prosperity, Solomon's annual income was 666 talents of gold, not including the revenues from merchants and traders and from all the Arabian kings and governors of the land (1 Kings 10:14-15). Six hundred sixty-six talents are about 22 tons of gold per year! King Omri of the Northern Kingdom bought the hill of Samaria on which he built his capital for two silver talents (1 Kings 16:24). And the Egyptian pharaoh Necho made the nation of Judah pay 100 silver talents and one gold talent after he had conquered her (2 Kings 23:33). This last figure indicates that gold was much scarcer and also much more valuable than silver.

The shekel, mina, and talent were international weights used in many different countries. We know they were used in Babylon from the famous story of the writing on the wall in the book of Daniel. The hand wrote, "Mene, Mene, tekel, parsin" (Daniel 5:25). The literal meaning of the words was clear to everyone present: "mina, mina, shekel, half-shekel." But, only the prophet Daniel could interpret what these words for money meant in this context: Just as minas and shekels must be weighed to find out if they are of proper weight, so God had weighed King Belshazzar and found him far short of God's expectations for a ruler. As a result, God had numbered his days and would divide his kingdom (vv. 26-28).

By the time of the return under Ezra and Nehemiah, a new system of money had come into use. The Persians and Greeks had begun using coins called drachmas (Nehemiah 7:70). Yet, this passage from Nehemiah indicates that valuable goods such as bowls and garments were still considered important commodities of wealth as well.

C. NEW TESTAMENT MONEY

The money used in New Testament times was more standardized and less diverse, partly because this period lasted only a century. Yet the New Testament monetary system was complicated by the fact that several systems of money were being used at the same time. Rome only allowed the Herods to issue bronze coins. The nearby cities of Antioch and Tyre were allowed to mint coins as well, while the Roman government issued copper, silver, and gold coins for use throughout the Roman Empire. By the time of Jesus's ministry, such coins were the most common type of money used for buying and selling, though bartering was still practiced as well.

The most common coin during New Testament times was the Roman silver denarius. The parable of the workers indicated that one denarius was the equivalent of a normal day's wages for a common laborer (Matthew 20:2 ff.). The denarius was a Roman coin which had the emperor's image and inscription on it (Matthew 22:19-21). Judas claimed that the perfume which the sinful woman poured on Jesus could have been sold for 300 denarii, which amounted to a year's wages (Mark 14:5).

The drachma, a Greek coin, was worth about the same as a denarius. The woman in Jesus's parable lost one of her ten drachmas and swept her house until she found it (Luke 15:8-9). Two drachma (or *didrachma*) coins and four drachma (or *tetradrachma*) coins were also in circulation. Jesus miraculously provided such a tetradrachma (called a *stater* in Latin) to pay the two-drachma temple tax for himself and Peter (Matthew 17:27).

Larger gold coins were also in use at this time. One mina was equal to one hundred denarii (remember one denarius was an average day's wage) in New Testament times, and ten minas still equaled one talent. Jesus spoke of talents in one of his parables, although recent translations often translate the word for talent as "bag of gold" (Matthew 25:15ff). A talent was an enormous amount of money, half a lifetime's earnings.

On the other hand, there were also copper coins worth much less than the denarius. And the Jews had an even smaller copper coin called the *lepton.* Two leptons equaled one *quadrans* or 1/16 of a denarius. A poor widow put two leptons into the collection box at the temple while Jesus was watching (Mark 12:42). In speaking of the punishment that a hypocrite will receive, Jesus says they will be cast into prison and not released until they pay every single lepton (Luke 12:59).

2 leptons = 1 quadrans

16 quadrans = 1 denarius/drachma (a day's wages)

4 denarii = l stater

100 denarii = 1 mina

60 minas = 1 talent

Chart 13. New Testament Monetary Units

That several different systems of coins were being used in New Testament times is seen clearly by references to the money changers in the temple. Jesus chased the money changers from the temple twice, once at the beginning of his ministry (John 2:14-16) and again at the end of his ministry (Matthew 21:12-13). Slightly later Jewish writings state that Jews coming to the temple were only permitted to pay their temple tax with shekels of Tyre.

The shekels of Tyre were acceptable not only because they had no human image on them but also because they had very high silver content at a time when other coins contained less and less silver due to inflation. The moneychangers took other coins in exchange and made a hefty profit. It was the greed of the money changers and the greed of the priests who had instituted this rule that made Jesus so angry.

LESSON 30:
NEW TESTAMENT GALILEE

Galilee was where Jesus grew up and where he spent his youth and much of his ministry. Many of his early disciples were also Galileans. In this lesson, we will take a close look at the geography, cities, and people of Galilee during the New Testament period.

At the end of this lesson you should be able to:

* *The rulers of Galilee during late Old Testament and New Testament times;*
* *List the four natural boundaries of Galilee;*
* *Summarize the reputation of Galileans during New Testament times;*
* *Name the two most common Galilean occupations;*
* *Give the approximate size of the Sea of Galilee; and*
* *Describe the importance of nine Galilean cities in the New Testament.*

A. THE LAND OF GALILEE

The northern part of the land of Israel was referred to as Galilee already in Old Testament times. Although Moses does not use that name, Joshua speaks of a city "in Galilee" (Joshua 20:7). This name comes from a Hebrew word meaning ring or circle. God gave the area of Galilee to the tribes of Asher, Naphtali, Zebulun, and Issachar. Isaiah uttered a prediction about this region when he said: "In the past God humbled the land of Zebulun and the land of Naphtali, but in the future he will honor Galilee of the nations by the Way of the Sea, beyond the Jordan" (9:1). In the following verses, Isaiah went on to prophecy about the work of Jesus, much of which later took place in Galilee.

For the most part, the borders of Galilee followed the geographical boundaries which we have put on our Bible map. On the west, Galilee extended to the Mediterranean Sea, and on the east, to the Jordan River. Its southern boundary was the southern edge of the Jezreel Valley, and it extended to the north as far as the mountains of Lebanon.

Geographically, Galilee was divided into two parts. Upper Galilee extended from the top of the Sea of Galilee northwards and is full of high hills and mountains. Lower Galilee was the area of lower hills west of the Sea of Galilee and included the Jezreel Valley. The Sea of Galilee (called the Sea of Kinnereth in Old Testament times, Joshua 11:2) was also referred to as the Lake of Gennesaret (Luke 5:1) and the Sea of Tiberias (John 6:1) in the New Testament. During the New Testament period, however, the state of Galilee did not contain all of this natural area. The Romans made the entire sea coast and part of Upper Galilee into the separate kingdom of Trachonitis (Matthew 4:12-16). Therefore, New Testament Galilee was quite small.

B. THE PEOPLE OF GALILEE

When the Assyrians destroyed the Northern Kingdom in 722 B.C., many of the leaders of Israel were carried into captivity never to return. Some commoners among the Jews remained in Galilee, but from this time on Galilee gradually became a Gentile region. Even after the return from exile, Gentiles continued to control much of the area. The Jews only regained control over portions of Galilee during the days of the Hasmoneans (see Lesson 26). During New Testament times it was surrounded by Gentile countries on three sides, and non-Jews continued to settle within its borders. Thus, as Isaiah had prophesied, it became known as "Galilee of the Gentiles" (Matthew 4:15).

Under Herod the Great, Galilee was united with Judea. During the life of Jesus, Herod Antipas ruled Galilee. His capital was first at Sepphoris, but then he built a new capital on the Sea of Galilee. He named it Tiberias in honor of the Roman emperor who ruled from A.D. 14 to 37.

Because of its long history under Gentile rule and influence, Galilee was not well respected by the Jews of Judea. Many Judean Jews viewed Galilean Jews as less faithful to Jewish customs, uneducated, and uncultured (Acts 2:7). They also spoke with a different accent from the Judean Jews (Luke 22:59). Yet Galilee also had a reputation for its fierce Jewish nationalism, and the Galileans frequently led the way in anti-Roman rebellions (Acts 5:37).

The land of Galilee was very fertile and well-watered. As a result, farming was the main occupation in Galilee. The Sea of Galilee was the setting for the other important occupation—fishing. Though some Galileans were well-off (Luke 12:13 ff.), most were poor. Fish and bread were the most common foods (John 6:9). The small farms, the open fields, the flowers and animals of Galilee provided Jesus with the illustrations for his preaching and parables. When God chose Galilee as the earthly home for his Son Jesus, he picked a humble area, yet one that contained many faithful followers of the true God.

C. THE CITIES OF CENTRAL GALILEE

The cities of Galilee mentioned in the Gospels are almost entirely in the region of Lower Galilee. The only city in Upper Galilee to which the Bible explicitly says Jesus went is Caesarea Philippi, a city near the source of the Jordan River. Most of Jesus's time was spent in Lower Galilee. We begin with Jesus's home town.

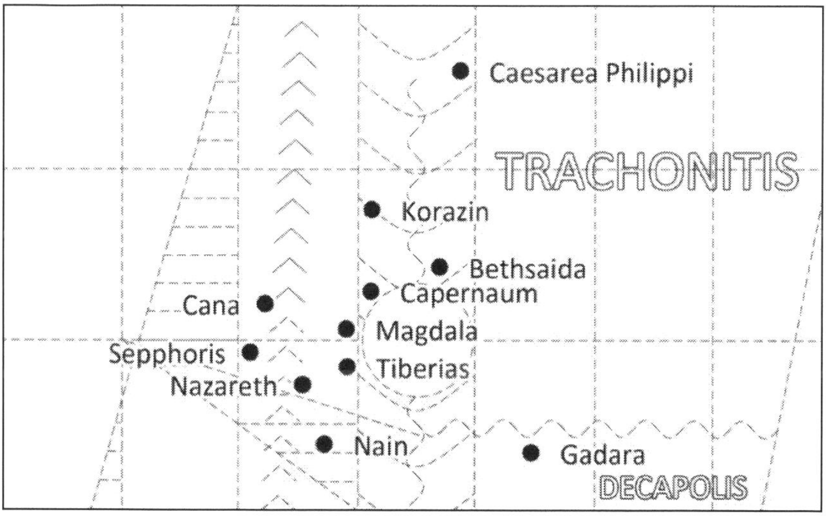

Map 23. Galilee in the New Testament

1) NAZARETH: Nazareth is most famous for being the boyhood home of Jesus. Joseph lived in Nazareth before the birth of Jesus (Luke 2:4), and Jesus's family returned there after their stay in Egypt. This small village was located on the northern edge of the Jezreel Valley, near the watershed, about 16 miles (25 kilometers) west of the bottom of the Sea of Galilee. None of the major roads or trading routes passed through or near Nazareth, so it was probably a rather quiet town in Jesus's day. Jesus preached one of his first sermons in Nazareth (Luke 4:16 ff.). Sadly, the Savior was rejected by his own townsmen. They even tried to kill him when

he claimed to be the fulfillment of the promised Messiah (vv. 28-30). After this event, Jesus made Capernaum (on the Sea of Galilee) his main home and the base from which he conducted his ministry. Yet, as had been prophesied, Jesus was still called a Nazarene (Matthew 2:23; Mark 16:6). Because of Jesus's connection to Nazareth, early Christians were also sometimes called Nazarenes (Acts 24:5).

2) SEPPHORIS: Sepphoris or Diocaesarea is not mentioned in the New Testament. It was a new city which Herod the Great built just four miles (6 kilometers) northwest of Nazareth. From there, Herod Antipas ruled Galilee during Jesus's boyhood. Sepphoris had Roman temples, theaters, and other Hellenistic buildings. Thus, Jesus grew up very close to one of the largest and most important cities of Lower Galilee, the seat of the government which ruled for Rome. Many Jews avoided Sepphoris because of its Hellenistic lifestyle.

3) CANA: Cana was located about 5 miles (8 kilometers) northeast of Nazareth. This village was the site of Jesus's first public miracle—turning water into wine at a wedding feast (John 2:1-11). During a second visit to Cana, Jesus healed the son of a royal official (John 4:46 ff.). Jesus's disciple Nathanael came from Cana (John 21:2).

4) NAIN: About 9 miles (14 kilometers) south of Nazareth was the town of Nain. It was built on the northern slopes of the hill of Moreh (Judges 7:1 ff.), a hill in the middle of the Jezreel Valley. When passing through this town, Jesus raised a widow's only son from the dead (Luke 7:11 ff.).

D. CITIES AROUND THE SEA OF GALILEE

Jesus spent much of his early ministry near the beautiful Sea of Galilee. This large body of water is located in the Rift Valley and was a well-known commercial fishing area. But even the local sailors feared its sudden and ferocious storms (Luke 8:22 ff.). Five cities along the shore of the Sea of Galilee are of special interest:

1) CAPERNAUM: The city of Capernaum was located on the northwest shore of the Sea of Galilee. As mentioned before, Jesus made his Galilean headquarters there. On the shores nearby, Jesus called his first disciples—Simon, Andrew, James, and John (Mark 1:16-20). Jesus also preached in the synagogue of Capernaum and cast out an evil spirit there (vv. 21-28). He healed the mother-in-law of Peter in Capernaum (Matthew

8:14-15). There Jesus met a centurion who had great faith (Matthew 8:5 ff.), and it was near the beach in Capernaum that Jesus performed the miracle in which Peter found money for the temple tax in the mouth of a fish (Matthew 17:24 ff.). In the nearby countryside, Jesus fed the 5000 (Mark 6) and preached his famous Sermon on the Mount (Matthew 5-7). Thus, this busy little fishing town became the center of Jesus's Galilean ministry.

2) BETHSAIDA: Bethsaida was a fishing town a few miles northeast of Capernaum, where the Jordan River enters the northern end of the Sea of Galilee. It was the hometown of Philip, as well as of Simon Peter and Andrew (John 1:44). Jesus's disciples were on their way to Bethsaida when Jesus came to them walking on the waters of the Sea of Galilee (Mark 6:45). There, Jesus restored sight to a blind man (Mark 8:22 ff.). Yet despite all the preaching and the miracles which Jesus did in this area, most of the local people still did not accept Jesus as the Christ. Instead, Jesus condemned the people of Bethsaida for their impenitence and rejection (Matthew 11:21).

3) KORAZIN: Korazin (or Chorazin) is located about two and a half miles (4 kilometers) north of the Sea of Galilee, beyond the cities of Capernaum and Bethsaida. The people of Korazin also heard Jesus's preaching and saw his miracles, yet rejected him. Today, the city is only a heap of ruins, a sign of the fulfillment of Jesus's judgment (Matthew 11:21).

4) MAGDALA: This village was south of Capernaum, along the west coast of the Sea of Galilee. It was the home of Mary Magdalene, the faithful follower of Jesus from whom he cast seven demons (Luke 8:2). The village of Magadan where Jesus spent some time preaching is probably another spelling for Magdala (Matthew 15:39), but we are not completely sure. Mark calls this area Dalmanutha (8:10).

5) TIBERIAS: Herod Antipas built the city of Tiberias to be his new capital. It was located south of Magdala on the southeastern shore of the Sea of Galilee. Herod named the city in honor of Tiberius, the Roman emperor at that time (ruled from A.D. 14-37). The city was begun in A.D. 20, and soon it had a population of 15,000 or more. At first, the Jewish population of the city was very small since many Jews avoided living among the "unclean" Gentiles. We have no record that Jesus visited the city. After the destruction of Jerusalem in A.D. 70, Tiberias became one of the religious centers of the Jewish religion.

In New Testament times, the area to the east of the Sea of Galilee was known as the Decapolis, a federation of ten cities. The most famous of the ten cities was Damascus, about 65 miles (100 kilometers) northwest of the Sea of Galilee. Damascus was the city to which Saul (soon to be the Apostle Paul) was headed to persecute Christians when Jesus appeared to him and converted him (Acts 9). Jesus also briefly visited the region of Gadara, another of the ten cities of the Decapolis. There he cast out a legion of demons (Matthew 8:28ff.). But it was the Sea of Galilee, and the hills, cities, and people of Galilee which were the setting for much of the ministry of Jesus the Nazarene.

LESSON 31:
JUDEA AND JERUSALEM IN THE TIME OF JESUS

Although Jesus spent most of his public ministry in Galilee, he also traveled to Judea and had followers among the Judeans. However, only a few Judean cities are mentioned by name in the Gospels. The most important, of course, was Jerusalem. In this lesson, we will take a look at Judea during Christ's time, and especially New Testament Jerusalem.

At the end of this lesson you should be able to:

* *Give the New Testament importance of these places in Judea: Bethlehem, Bethany, Emmaus, Jericho, the wilderness of Judea, and Jerusalem;*
* *Give the location and importance of the following sites in Jerusalem: the Antonia fortress, Herod's palace, the Hasmonean palace, Golgotha, Gethsemane, the Pool of Bethesda, the Pool of Siloam, and the temple; and*
* *Give the location and importance of these places within the temple area: the highest point of the temple (pinnacle), Solomon's Colonnade, Court of the Gentiles, Court of the Women, Court of Israel, Holy Place, and Most Holy Place.*

A. JESUS AND JUDEA

Judea was the center of government, religion, and trade for all of Israel at the time our Savior lived on earth. And Jerusalem was the center of Judea. Even though Jesus spent a good deal of time in and around the city of Jerusalem, the Gospel writers tell us relatively little about his preaching activity in other parts of Judea. Several cities and areas of Judea did play a role, however, in the life and work of Jesus (see Map 24):

1) BETHLEHEM: The small town of Bethlehem, located about six miles (10 kilometers) south of Jerusalem, had been the hometown of King David. It was also the birthplace of Jesus (Luke 2:5-7), which fulfilled God's prophecy through the prophet Micah (5:2). After Mary and Joseph fled with the baby Jesus to Egypt, Herod murdered the children of Bethlehem

(Matthew 2:13-18). Bethlehem played no further role in New Testament history.

2) THE WILDERNESS OF JUDEA: The wilderness of Judea was the strip of hot, sandy, empty land between the watershed and the western shore of the Dead Sea. It was the setting for several New Testament events. John the Baptist lived and preached there during his ministry. The area had few inhabitants, and little grew there, so John ate grasshoppers and wild honey (Matthew 3:1 ff.). Jesus also spent forty days and nights in the wilderness of Judea, where he was tempted by the devil (Mark 1:12-13). The story of the Good Samaritan was also set here (Luke 10:25-37).

3) JERICHO: The city of Jericho was located just north of the Dead Sea on the western side of the Jordan River. For much of the Old Testament period, Jericho was the main city of southern Palestine. When Israel returned from Egypt and entered the Promised Land, the heavily fortified and strategically located city of Jericho was the first city God enabled them to conquer (Joshua 6). Although it was completely destroyed at that time, Jericho was later rebuilt (1 Kings 16:34). By New Testament times, Jericho was again one of the most important cities of Judea. Because of the mutual animosity between Jews and Samaritans, Jews did not like to travel through Samaria. So when traveling from Galilee to Judea (or vice versa), they would often down into the Rift Valley, travel along the Jordan River, and then "go up" again through Jericho to Jerusalem. Jesus was traveling this way when he passed through Jericho and healed the blind beggar (Luke 18:35 ff.). Jesus also met the tax collector Zacchaeus while traveling through Jericho (19:1 ff.). Herod the Great had built a whole new city of Jericho to serve as one of his winter palaces. He built it a short way from the ancient tell of Jericho. Thus, both a "new Jericho" and an "old Jericho" existed in New Testament times.

4) BETHANY: The small village of Bethany is about one and a half miles (2.5 kilometers) east of Jerusalem. It was the home of Jesus's friends Mary, Martha, and Lazarus (John 11:1). Bethany and the nearby village of Bethphage were both located on the slopes of the Mount of Olives (Luke 19:29), along the road between Jerusalem and Jericho. Jesus passed through both as he approached Jerusalem on Palm Sunday. After he taught all day in Jerusalem, Jesus sometimes spent the night in Bethany, perhaps at the home of Mary and Martha (Matthew 21:17). Simon the Leper, another resident of Bethany, also showed hospitality to Jesus (Matthew 26:6).

5) EMMAUS: Emmaus was a small village about seven miles (11 kilometers) northwest of Jerusalem. Its only appearance in the New Testament is in Luke 24 when Jesus spoke with two disciples as they traveled to Emmaus on Easter Sunday.

B. JERUSALEM

The city of Jerusalem in Jesus's day was a large and beautiful city. Herod the Great had done much to expand and beautify the city during his long rule. Because the modern city of Jerusalem still covers the area of the ancient city, archaeologists have never been able to unearth much of ancient Jerusalem. Legends have been handed down about many of the famous Bible sites in Jerusalem, but often these suggested locations are unreliable.

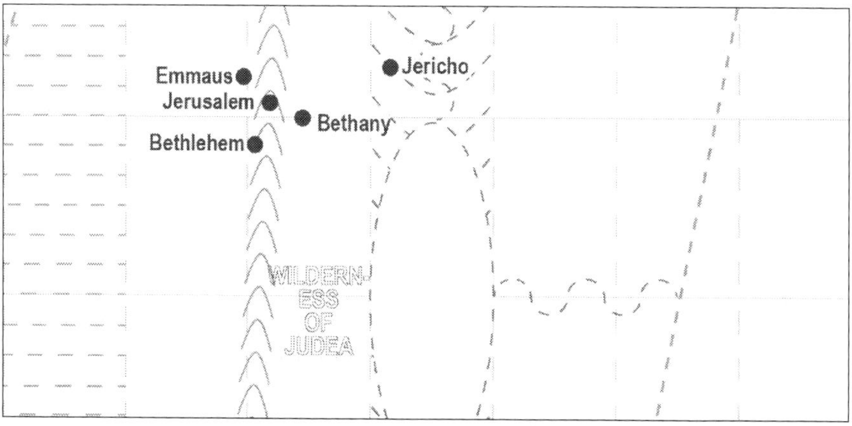

Map 24. Judea During Jesus's Time

In Lesson 18, we studied the Old Testament city of Jerusalem and its geography. Although the geography did not change, by New Testament times the city had expanded far beyond Mt. Zion and Mt. Moriah. The walls of Jerusalem then enclosed most of the area between the Hinnom and Kidron Valleys.

On the western side of the city was Herod's palace, which Herod the Great had built. Pontius Pilate and the other Roman governors made this palace their headquarters while visiting Jerusalem. About half-way between Herod's palace and the temple was the Hasmonean palace, built by the Hasmonean rulers. Herod Antipas probably stayed there when he visited Jerusalem, as he did during the trial of Jesus (Luke 23:7). At the northern end of the temple courts was the Roman fortress of Antonia, where the main body of Roman soldiers was stationed. The Gospels do not specifically tell us whether Jesus was tried by Pilate at the Antonia Fortress or at Herod's palace.

The garden of Gethsemane, where Jesus prayed the night he was arrested, was almost certainly at the foot of the Mount of Olives in the Kidron Valley, to the east of Jerusalem (John 18:1; Matthew 26:36). Archaeologists believe that the house of the high priest Caiaphas was in the southwestern section of the city. Jesus spent much of the night at that house before he was handed over to Pontius Pilate on Good Friday (Matthew 26:57; Luke 22:54). Jesus was then crucified on a hill known as Golgotha—the Hebrew word meaning "place of the skull" (Matthew 27:33; "Calvary" is the Latin word for skull). Golgotha was probably on the northwest side of the city "outside the city gate" (Hebrews 13:12).

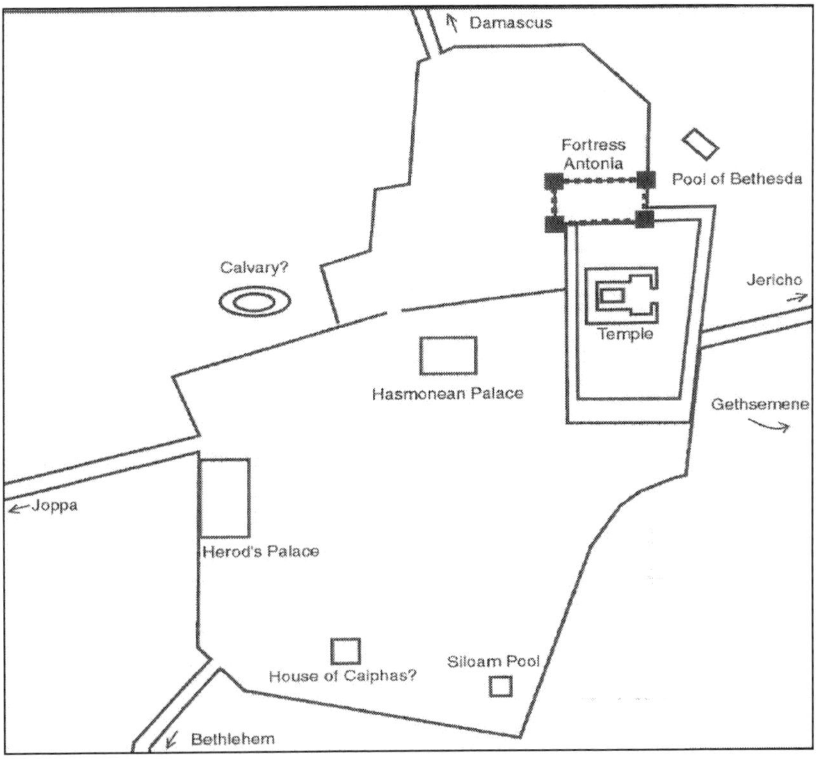

Map 25. Jerusalem in the New Testament Period

Jesus performed miracles involving two different pools in Jerusalem. Jesus put mud on the eyes of a blind man and told him to wash in the Pool of Siloam (John 9:7), the ancient pool at the southern edge of the city. At the Pool of Bethesda, located north of the temple, Jesus cured a man who had been disabled for 38 years (John 5:2). These, then, are the sites of Jerusalem that are mentioned in the New Testament.

C. THE TEMPLE OF HEROD

After the Israelites returned from their Babylonian captivity, they rebuilt the temple in Jerusalem, but it was smaller and less beautiful than the one Solomon had built. Herod the Great resolved to replace that small temple with a temple complex that was far larger and more grandiose than the temple of Solomon (see Chart 14). The entire project took over eighty years and was not completed until sixty years after Herod's death. He began about 20 B.C., and the temple was not completed until about A.D. 65, just a few years before it was completely destroyed by the Romans. The temple Herod built was a magnificent feat of engineering. While one of Jesus's own disciples was looking at the temple, he commented, "Look, what massive stones, what magnificent buildings!" (Mark 13:1).

Herod began the project by first enlarging Mt. Moriah. He built up its slopes and constructed a platform on top that was about 1600 feet (490 meters) long and 900 feet (275 meters) wide. Around the edge of the platform, he built large colonnades, roofed areas supported by rows of beautiful pillars. Facing the east and overlooking the Kidron Valley was Solomon's Colonnade. Jesus would frequently teach the people in the shade beneath this large covered walkway (John 10:23). After Jesus's ascension, the early Christians continued to meet there (Acts 5:12). The money changers sat with the buyers and sellers of animals in the southern colonnade (Mark 11:15-16). In that same area, the Sanhedrin had its normal place of meeting (Acts 5:21 ff.).

At the edge of the porch on the southeastern corner, there was a great drop straight down to the bottom of the Kidron Valley. This area was called the pinnacle or "highest point of the temple." The devil took Jesus there and told him to throw himself down (Matthew 4:5). At the opposite corner of the temple area, the northwestern corner, stood the Antonia fortress, where Roman soldiers watched over what went on in the temple courtyards.

Roman guards ran out from there to rescue Paul when he was being lynched by a mob (Acts 21:31-32).

Inside the porches were the temple courts. The largest area was called the Court of the Gentiles. This court was the only area a non-Jew could enter. Large signs warned that the punishment for going further was death. The Jews had a special temple guard, a group of Jewish policemen, who kept order and enforced proper behavior in the temple area (Acts 4:1). Jewish women were permitted to enter into the next court, called the Court of Women, but were allowed no further. The temple collection boxes were placed in this area (Mark 12:41). Jewish men were allowed to go still further into what was known as the Court of Israel. Beyond this was the Court of the Priests in which the altar for burnt offerings and the large bowl for the priests to wash were located.

Finally, inside the Court of the Priests was the temple building itself. This was made up of the same two parts as the earlier Tent of Meeting—the Holy Place and the Most Holy Place. Only the priests were allowed to enter the first room, the Holy Place. Separated by a thick curtain from the Holy Place was the Most Holy Place, where the Ark of the Covenant was kept. This curtain between these two areas was ripped in half from top to bottom at the time of Jesus's death, showing that Jesus's death had reopened access to God's presence for all believers (Matthew 27:51). When the Romans destroyed the temple and much of Jerusalem in A.D. 69-70 they took the Ark and the other temple furniture back to Rome as a sign of their victory over the Jews.

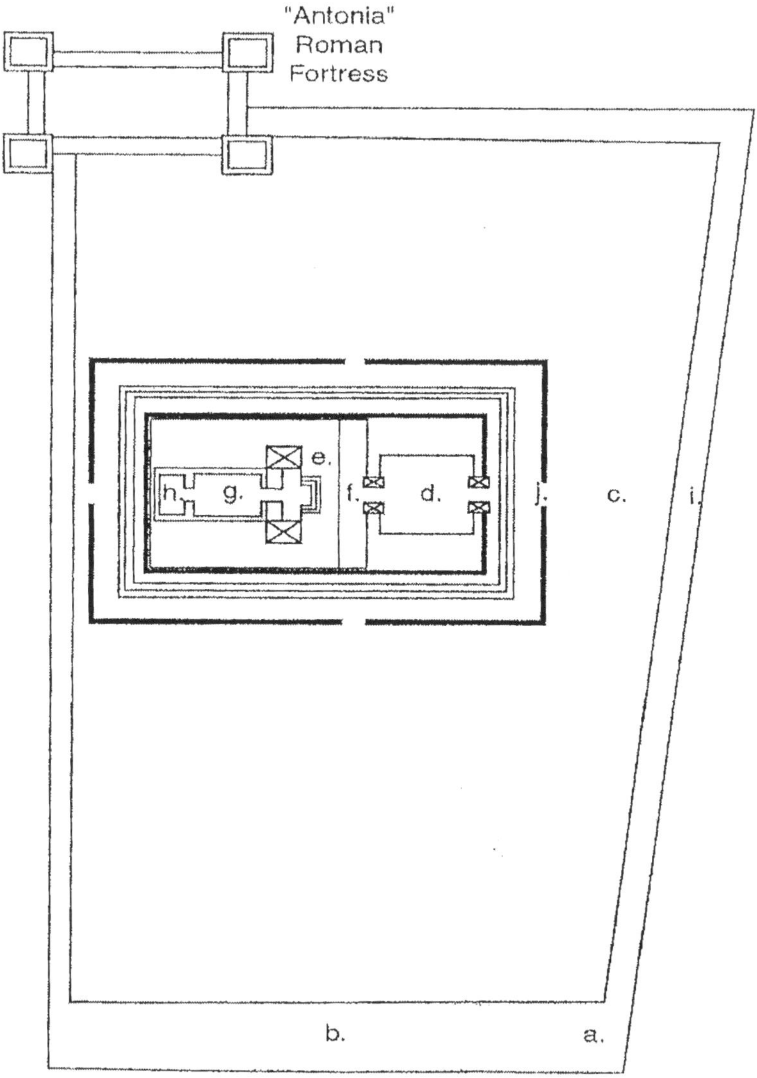

Chart 14: Herod's Temple

a. Highest point of the temple
b. Place of the money-changers and Sanhedrin
c. Court of the Gentiles
d. Court of the Women
e. Court of the Priests
f. Court of Israel
g. Holy Place
h. Most Holy Place
i. Solomon's Colonnade

LESSON 32: HOUSES, CITIES, AND JEWISH HOME LIFE

The basic way of life among the Hebrew people changed very little from the time of the Exodus until New Testament times. This is clearly shown in Jewish houses and cities. In this lesson, we will look at the methods of building houses as well as the basic parts and floor plans of houses in Bible times. We will also discuss the construction of cities as well as life in a normal Jewish home.

At the end of this lesson you should be able to:

* *Describe the three types of Jewish houses and who lived in each;*
* *Describe the use and significance of the foundation, the cornerstone, the door and windows, the roof, and an upper room;*
* *Tell where cities were usually built and describe some of the basic characteristics of their layouts;*
* *Describe what took place at the gates of a city; and*
* *Tell how addresses were given in cities during Bible times.*

A. TYPICAL HOUSE

The typical Jewish house usually consisted of just a single room. That room was rectangular or square in shape and had a flat roof. Most houses were built of sun-dried bricks (Job 4:19; Isaiah 9:10). Since Israel was a land full of rocks and stones, however, houses were also built of stones held together by mortar. Only the rich were able to afford to build houses with cut and trimmed stones. Sometimes the bricks and stones might be covered with whitewash after they were completed (Ezekiel 13:10 ff.). Because the land of Israel was so rocky, Jews had no excuse for not building their houses on solid foundations (Matthew 7:24 ff.). Most of the time there was no need to dig a foundation below ground as we do today. The builder started construction by laying a "cornerstone" which was as straight and square as possible. The edges of this stone were used to align the rest of the building—horizontally along the base of the walls and vertically up to the roof. Thus, the cornerstone was crucial in laying out a strong and straight building. Paul uses this picture as an illustration of the importance of aligning Christ's church with Christ himself and his teaching (Ephesians 2:20). The builder would suspend a plumb line (a string with a weight on the end) from above to ensure the walls were being built straight (Amos 7:7-8). At times, wood might also be used to strengthen the walls. The only breaks in the walls were for the door and for whatever windows were desired.

The door was usually two or three feet (1 meter) wide and about six feet (1.8 meters) high. A large flat stone was usually placed at the bottom of the door to prevent water or rodents from entering. A very shallow ditch dug just outside the door further assisted in keeping out rainwater. A doorpost, a solid piece of wood or stone, was placed on each side of the opening. Another stone or piece of wood was then placed across the top of the doorposts to form the lintel, the top of the door frame. Exodus 12:22 tells how the blood of the Passover Lamb was spread on the lintel and doorposts. The door was normally made of wood and was fastened in place by hinges. The occupants could bar or bolt the door from within to keep out intruders (2 Samuel 13:17). Some people had locks and keys already by the time of the judges (Judges 3:23-25).

The windows were built in the same manner as the doors, using wood or stone as frames. Windows were placed quite high up in order to make it difficult for thieves to enter the house (Joel 2:9). The windows usually remained open to the air. A cloth or carpet could be suspended over the opening during bad weather. Other houses had a lattice to keep out the direct sunlight and wind. This allowed people inside to look out while preventing people outside from looking in (Judges 5:28; Proverbs 7:6). Lower windows needed shutters or bars for protection. If a house had several stories, the upper windows might have a lattice covering them or merely be left open. This latter practice could prove dangerous (2 Kings 1:2; Acts 20:9-12), but could also prove to be very useful (Joshua 2:15; 2 Corinthians 11:33).

The roof of the house was usually flat, though in New Testament times, some Jews began building houses with a kind of domed roof. The flat roof was made of branches and other pieces of wood or tile which were laid across the tops of the walls. Then they added straw, dirt, and clay and packed them down tightly. The process was then repeated several more times until the roof was about one foot (1/3 meter) thick. At times grass would begin to grow out of this earthen roof, but since it was exposed to the hot sun, it often quickly died (Psalm 129:6). After each major rainfall, the roof would have to be rolled and pressed so that it did not leak or weaken. An outside staircase led to the roof so that it could be properly maintained. During the dry season when the weather was hot, people might sleep on their roofs to take advantage of any cooler breeze.

B. USING A HOUSE

Ancient houses were not primarily built to live in. Most of the time people lived outside. The house was a place for sleeping and a place for protection from the elements. As a result, the insides of most houses were not very elaborate. One room served all purposes. Privacy in a house was almost impossible. Everyone slept together in the same room (Luke 11:7). The most common furnishings of the house were mats or a simple bed for sleeping, a few cooking posts, some lamps and perhaps a table. A lamp was placed so that it could throw light on the entire room. Such a lamp was often set on an overturned bushel or basket, but it would make no sense to put the lamp under the basket (Matthew 5:15).

The roof became a multi-purpose area. People dried and stored fruit and grain there (Joshua 2:6). Children at times played on the roof. Therefore, God had commanded that roofs must always have a low wall or parapet around their edges to prevent people from falling off (Deuteronomy 22:8). Important announcements in a village could be made from the roof of a centrally located house (Luke 12:3). People might also go to the rooftop to get a good view of what was happening nearby. In extreme cases, a hole could be opened in the roof to gain access to the inside of the house (Mark 2:4). However, many people went up to their roof as a simple way to get away from the noise and excitement of life in the city. Therefore, the roof was also an ideal place for prayer and meditation (Acts 10:9). Especially during the hottest times of the year, the roof would be a cool place to sleep or to relax in the early morning or evening (1 Samuel 9:26). Finally, in time of danger, the roof could be used as a place of safety (Joshua 2:6).

Houses often had an upper room built on a corner of the rooftop, accessed by the outside staircase. Jesus celebrated his final Passover meal with his disciples in such a room (Mark 14:15). It was also in an upper room that the early Christians continued to meet in Jerusalem after Jesus's resurrection (Acts 1:13). Elijah lived in such a room when staying with the widow of Zarephath (1 Kings 17:19). The Shunammite woman talked her husband into building this kind of room so Elisha could stay with them (2 Kings 4:10). Thus, we see that an upper room could serve as a storehouse, an extra room for an expanding family, or as a guest room. The book of Proverbs points out that the upper room "on the corner of the roof" might also be a place to get away from the frustrations of family life (21:9). Since the rooftop room was cooler during the summer, and the ground floor room was warmer during the winter, perhaps these two parts of the house are meant when we read of summer houses and winter houses--although wealthy Jews may often have owned two separate houses (Amos 3:15).

C. MORE ELABORATE HOMES

The wealthier people of Palestine often had more elaborate and larger homes. A middle-class merchant, for example, might build a wall around his home. The area inside the wall could then become a garden for growing vegetables, herbs, olive and fruit trees, or even a vineyard. The stairway to the roof could be put inside the garden, or even inside his home to give it additional privacy and security. A still more elaborate type of house was often built by the upper-class city dwellers. Often this took the form of a large square building. A large outside door led through a passageway to a center courtyard. A doorkeeper, usually a servant or slave, would receive visitors (Psalm 84:10; Acts 12:13). There might also be a well or pool in the center of the courtyard (2 Samuel 17:18-19). Well cared for trees would be found there to provide shade (Psalm 52:8; 92:13). At night, the people of the house or the servants might keep themselves warm around a fire in the courtyard (John 18:18). The rooms lining the four sides could be entered only from the courtyard.

This was a popular design among the Romans and became increasingly common and popular in Israel in New Testament times. The house itself was probably built of finely cut stones (Isaiah 9:10) and would be well furnished. More ornate wooden lattices might cover the windows allowing the people to look out while preventing outsiders from looking inside (Proverbs 7:6). Bathing might also be done in the courtyard (2 Samuel 11:2). Those hearing Jesus's story of the rich man at whose gates Lazarus begged would have pictured this type of house (Luke 16:19 ff.), and the high priest Caiaphas (John 18:13 ff.) undoubtedly had a similar house.

D. CITIES

The cities of ancient times were usually not well planned. They typically grew up naturally near sources of water in places that could be easily defended. Often, they were built on top of earlier cities creating tells (see Lesson 19). Because cities often developed naturally, they were frequently stretched out along main roads and paths. Houses were built next to one another without much planning, which often resulted in narrow and crooked streets. The larger cities were surrounded by thick walls to keep out enemies. The old men and leaders of the city often sat near the gate of the city to judge the cases which people brought to them (Deuteronomy 22:24; Ruth 4:11).

By New Testament times, the leaders of the people put more thought into planning out their cities. Herod the Great, already noted for his building program, laid out and built several new cities. When cities were able to be planned and built in this way, the streets would be more even and straight, the houses spread further apart, and specific areas set aside for business, trade and commerce. Streets of larger cities were often given names, similar to today. An address included the name of the man and the street on which he lived. When Saul was in Damascus, the Lord directed him to go to "the house of Judas on Straight Street" (Acts 9:11). Or else, the address was simply a reference to a nearby important feature within the city. An angel told Cornelius to send his men to Joppa to find Peter at the house of "Simon the tanner, who lives by the sea" (Acts 10:32). By New Testament times, Hellenized cities also provided theaters, baths, stadiums, and other public benefits for their people.

E. JEWISH HOME LIFE

Jewish home life was somewhat different from that of its surrounding Near Eastern neighbors. Women were generally held in higher regard. Children—both males and females—were treated as special gifts from God. Husbands and wives were viewed much more as a unit. As in many ancient cultures, the extended family often lived together or close to one another. Together they joined in household chores, worked in the fields, and brought up children.

Although some rich people practiced polygamy in earlier times, the typical Jewish home consisted of one husband and one wife. The Jews took their religious laws about marital life and the upbringing of children very seriously. The father was responsible for providing leadership in the home, but the influence of a God-fearing mother was also important. Some boys learned to read and write and memorize sections of the Scriptures at the synagogue. By the age of twelve, they were considered old enough to assume some adult responsibilities, such as attending the festivals in Jerusalem (Luke 2:42). By their mid-teens, they were permitted to marry, purchase property, and go to war. However, they were not considered full men until they were 30 years old (Luke 3:23).

The ideals of family life were clearly taught in the Old Testament and New Testament writings, even though the Jews often did not consistently put them into practice. Jewish home life was to be centered around Jewish religious observances. Children were to grow up seeing their parents regularly engaged in prayer and carrying out ritual washings. The observance of the Sabbath and its accompanying rituals were also of great importance to the family. The main religious festivals were to be carefully explained to every child (Exodus 12:24-28), and participation in synagogue worship and at the major festivals was expected. A warmth and tenderness within the family would be a direct outgrowth of the practice of these religious principles. Faithfulness to one's spouse, devotion to one's children and bringing them up in a God-pleasing atmosphere were of the highest importance. Christian homes today still strive to achieve these same ideals with Christ and his gospel as their example and motivation.

LESSON 33:
JOBS AND OCCUPATIONS

Paul wrote to early Christians saying, "If a man will not work, he should not eat" (2 Thessalonians 3:8). Work has always been necessary to provide food and clothing for survival. But just what kind of work did people do in Bible times? What were their jobs and occupations? In this lesson, we will take a look at some of the important occupations of people in Bible times.

At the end of this lesson you should be able to:

* *Name and describe the four food-producing occupations in Bible times;*
* *Describe metal working, pottery making, weaving, and carpentry and the main tools used in each;*
* *List two Bible chapters dealing with sheep and shepherds; and*
* *Explain the spiritual truths portrayed in the Bible through the pictures of the shepherd, the potter, and the fisherman.*

A. FARMING AND AGRICULTURE

Everyone needs food, and farmers grow most of it. Thus, farmers and the development of agriculture are important aspects in the advancement of a civilization. In Lesson 15, we studied the three main crops grown in Israel – grain, grapes, and olives. In that lesson, we also explained how these crops were grown, harvested, and made ready for consumption. Therefore, in this lesson we will not discuss farming in detail. We should note, however, that the vast majority of people in ancient Israel lived in rural areas and were subsistence farmers—people who grew enough food for themselves and their own families, and hopefully had enough extra left over to sell or barter for clothing and other essentials that they did not make for themselves. A small percentage of farmers became wealthy. They not only could hire laborers to work in the fields for them (Matthew 20:1-16) but also built barns to store all their produce (Luke 12:18).

B. HUNTING AND FISHING

Hunting and fishing were other important ways of obtaining food during Bible times. Hunting was one of the earliest ways that people secured food. God told Noah after the flood that "everything that lives and moves will be food for you" (Genesis 9:3). But since many of the animals that were good to eat moved rapidly and had a natural fear of humans, learning hunting skills was necessary. Esau is one of many Bible characters who were skillful hunters (Genesis 25:27). The land of Israel was originally full of wild animals (Exodus 23:29). Deer, gazelles, roebucks, and various kinds of birds were among the animals hunted by Israel's early inhabitants (1 Kings 4:23). Swords, spears, darts, javelins, arrows, slings, clubs, and lances were among the weapons that hunters used to kill their prey (Job 41:26-29). Hunters also dug pits for the animals to fall into (Ezekiel 19:4), laid nets to trap them (Isaiah 51:20), and prepared snares for them (Psalm 124:7). The Jordan Valley north of the Dead Sea was an especially good area for hunting (Jeremiah 49:19). During the time of the year when the Rift Valley flooded, hunters drove animals from there into the wilderness of Judea where they could more easily be killed.

Fishing was also a common way of obtaining food. Fish was one of the favorite foods of the Israelites during their stay in Egypt (Numbers 11:5), and it remained a favorite for many centuries afterwards. However, in the land of Israel there are few places to fish. Only the Jordan River and the Sea of Galilee were good fishing areas for the Jews. The ancient Jews used several methods of fishing. Sometimes they used hooks (Isaiah 19:8; Matthew 17:27); sometimes they speared the fish (Job 41:7); but most commonly of all, they threw out nets to catch the fish. At times smaller hand nets were used (Mark 1:16-17), but more often larger nets were dragged along the bottom of the Sea of Galilee, and the nets full of fish were then brought to shore (Matthew 13:47-48) or taken into the fishermen's boats (Luke 5:6-7). Night time was usually the best time for fishing (Luke 5:5). Jesus chose several fishermen as his disciples and used fishing as a picture of the mission activity he wished his followers to engage in (Luke 5:10).

C. HERDING

Herding animals such as sheep, goats, cattle, and the like was a third way of securing food. We have already studied how herding played an integral role in the lives of the early patriarchs (Lesson 8). Throughout the history of Israel, maintaining herds remained important, which is shown by the large sizes of the herds people possessed. Job, for example, had 14,000 sheep under his care (Job 42:12). When the temple of Solomon was dedicated, 22,000 cattle and 120,000 sheep were sacrificed (1 Kings 8:63). Cattle, sheep, and other domesticated livestock were an integral part of the life and food supply of God's people.

Sheep and goats were the main form of livestock among the common people of Israel. Wool was used to make clothing, and sheepskin to make coats (Leviticus 13:47); they ate the meat or used it for sacrifices (2 Samuel 12:4); the horns were used for carrying water and oil (1 Samuel 16:1), to sound announcements (Psalm 81:3), or just for decoration. They used the milk of sheep and goats as well as cows for drinking and for making cheese (2 Samuel 17:28-29), as well as in cooking (Exodus 23:19).

The people who took care of a flock of sheep and goats were known as shepherds. Often younger boys were given this job (1 Samuel 16:11), but it involved great responsibilities, especially if the flock was large. A shepherd's job never ended, for he had to care for his flock night and day. He might carry a small bag or pouch (1 Samuel 17:40). He also carried a club, rod, or sling to protect the sheep from thieves and wild animals. Many shepherds also carried a staff both to assist in walking as well as a weapon. He used the staff also to guide the sheep and keep them together (Psalm 23:4). During the long hours spent with only their sheep for company, some shepherds sang or played music to occupy their time and to calm their flocks. The shepherd boy David became a famous musician (1 Samuel 16:18) and songwriter, and he composed many of the Psalms.

The shepherd bore the responsibility for the health and safety of his sheep. He had to make sure that the flock had enough grass to feed on and water to drink. He had to protect them from wild animals and from thieves, even risking his own life to keep them safe (John 10:15). At times, shepherds would use caves or sheep pens to protect their flocks (John 10:1).

While rich people could hire shepherds to take care of their flocks (John 10:13), good shepherds would always know their sheep personally. Often, several herds would be mixed together while drinking (Genesis 29:1-3), but a good shepherd always knew which sheep were his, just as his sheep recognized and followed only his voice (John 10:4-5). That is one reason why the Bible uses the shepherd and his sheep as a picture of God and his people (Psalm 23; John 10).

Except for city dwellers, almost all Israelites did some combination of farming, herding, and hunting or fishing. Within each family, certain people also took responsibility for other necessary activities—building and repairing the house, making implements, treating the sick, etc. As time went on, some of these part-time specialists became full-time craftsmen and professionals. In the remaining part of this lesson, we will look at some of the most common crafts and specialized occupations of the people who lived in Israelite towns and cities.

D. METAL WORKING

The profession of metalworking went back to the days of Tubal-Cain (Genesis 4:22) and was always vital to the progress of civilization. Until about the time of King David, the metal most commonly used by the Israelites was bronze. Iron began to be used more commonly about 1000 B.C. People mined iron and copper in the land of Israel already at the time of the Exodus from Egypt (Deuteronomy 8:9). Metals were also imported from other lands (Ezekiel 27:12). The metalworker's main task was to form various metals into useful tools and articles needed for daily life

The first challenge for the metalworker (also called a blacksmith, Isaiah 54:16) was to refine and purify the metal (Malachi 3:3). He would use a bellows made of animal skins to blow air onto a fire (Jeremiah 6:29). The very high temperature softened the red-hot metal. The worker then took the metal from the fire with tongs and beat and shaped it with a hammer (Isaiah 44:12). This hammering took place on a very hard rock or piece of metal called an anvil (Isaiah 41:7). In this way the blacksmith formed and sharpened plows, blades, axes, and sickles (1 Samuel 13:19-21). Silversmiths and goldsmiths were special kind of metalworkers who made finely crafted and decorated objects, usually for the rich or the royal family (2 Samuel 8:10). Sad to say, metalworkers in Israel also engaged in idol making as well (Isaiah 40:19).

E. POTTERY MAKING

Almost all the cooking pots, dishes, lamps, and storage vessels in Bible times were pieces of earthenware pottery (baked clay). Such pottery was cheap to make and to buy (Leviticus 11:33). Pottery also broke very easily, and because of this was constantly being replaced. The potter dug his clay from a field, soaked it in water, and then pressed it with his feet until it was soft and workable (Nahum 3:14). After the potter had taken all the good clay from a field, the field was often turned into a garbage dump. The chief priests bought such a "potter's field" with the money Judas returned to them (Matthew 27:7).

The potter took the clay to his workshop where he formed a piece of clay into the shape he desired by using a potter's wheel (Jeremiah 18:1-6). The skill and creativity of each individual potter was later displayed by the beautiful and unique bowls and jars that he produced. This creative skill of the potter in fashioning the clay to suit his purposes is used several times in the Bible as an illustration of how the Lord shapes his people and makes them useful instruments for his work (Jeremiah 18:6; Isaiah 64:8; Romans 9:20-21).

F. WEAVING

Throughout Bible times, most people produced their own clothing at home. This involved first spinning cotton or flax into thread and then weaving that thread into cloth. Some people, however, made this their full-time occupation, and they produced cloth to sell locally and also for export. Proverbs 31:10-31 pictures a faithful wife at work. First she spun the thread. She did this by putting the cotton on a fork-like object called a distaff (v. 19), then spinning the cotton with her fingers and wrapping it around a spindle to keep it from becoming tangled (v. 19).

Next the thread was woven into cloth on a loom. Every woman had a small loom in her home. Delilah used this kind of loom to weave together Samson's hair (Judges 16:13-14). Professional weavers used larger looms. The loom consisted of a large wooden frame with numerous pegs on all four sides. Single pieces of thread were suspended across the loom from one side to another until the entire loom was filled. Then, another thread was fastened to a shuttle. The thread was woven in and out between the other horizontal threads from top to bottom. The weaver's beam was a thick wooden stick that was used to lift every other thread so that the shuttle could pass through very quickly (Job 7:6; Goliath's spear was compared to a weaver's beam in 1 Samuel 17:7). This was repeated until the cloth was finished after a long and laborious process. It was then cut from the loom and rolled up (Isaiah 38:12). It could then be stored until it was used.

To make the thread extremely white and attractive, the material was often bleached by a fuller. Modern translations call him a washerman or launderer. The launderer took clay from his launder's or washerman's field (Isaiah 7:3; 2 Kings 18:17) and soaked the material in a paste made from the clay. Then he placed the cloth out in the sun. After the cloth had dried, he rinsed it with water and dried it a second time. This could be repeated until the cloth became dazzling white (Mark 9:3).

Other pieces of cloth were dyed into a multitude of different colors. Animal skins were also dyed at times (Exodus 25:5). Different cities and areas became famous for specially colored dyes that they produced. Lydia, one of the first converts to Christianity in the city of Philippi, was a seller of the famous purple cloth from the city of Thyatira in Asia Minor (Acts 16:14).

G. CARPENTRY

In ancient times, much of Israel was forested. So naturally, many items for homes and businesses were made of wood. The crafting of wood products was the work of a carpenter. Trees were cut with axes. Care always had to be taken so that the ax head would not fly off of its handle and injure someone (Deuteronomy 19:5) or be lost (2 Kings 6:5-7). The wood was then transported to the carpenter's shop where the carpenter used a saw to cut it to the needed sizes (Isaiah 10:15). The carpenter then shaped the wood with a chisel (Jeremiah 10:3), drilled any needed holes with an awl (Exodus 21:6), and nailed pieces together using a hammer and nails or wooden pegs (Jeremiah 10:4).

Jesus's earthly father Joseph was a carpenter (Matthew 13:55). Most likely Joseph made simple farm tools or house furnishings. Jesus, like other boys, was taught his father's trade (Mark 6:3). Being a carpenter was a humble but honorable occupation.

H. OTHER OCCUPATIONS

We have discussed the occupations most often mentioned in the Scriptures. Other occupations, however, are referred to occasionally. Tanners and leather workers prepared skins and sewed them into clothing, bags, and other articles (Acts 10:6). Tentmakers like Paul and Aquila and his wife Priscilla (Acts 18:3) made tents, awnings, and shades from skins and cloth. Stone cutters prepared stones for use in building (1 Kings 5:18), and masons then used these stones in constructing various buildings (2 Chronicles 24:12). Bakers baked bread (Jeremiah 37:21). By New Testament times, professional doctors specialized in helping the sick and diseased (Mark 5:26). One such doctor was the Gospel writer Luke (Colossians 4:14). Merchants traveled throughout the Near East purchasing and exchanging goods (1 Kings 10:15), and others sold their goods in city shops and open-air markets (James 4:13). Chopping wood and carrying water were considered two of the lowest jobs (Deuteronomy 29:11). Bankers borrowed and lent money at interest (Matthew 25:27). Trusted men served as scribes, priests, and rabbis (teachers), but seldom were these full-time occupations. And, like today, many casual laborers waited to be hired and to earn some money doing whatever occasional work they could find (Matthew 20:12). A boy was usually trained in his father's occupation and then followed in that same type of work.

LESSON 34: BIRTHS, MARRIAGES, AND DEATHS

Every ethnic group has its own distinctive religious and social customs for the special occasions that take place in family life. Especially important are those that take place at the four major changes in life—birth, becoming an adult, marriage, and death. In this lesson, we will look at the biblical customs associated with these events.

At the end of this lesson you should be able to:

* *Describe the care given a newborn baby, what took place on the eighth day for every male, and what the mother had to do shortly afterwards;*
* *Tell when a Jewish child was considered an adult in Jewish society;*
* *Describe who arranged a betrothal and what it involved;*
* *Identify the two main parts of a Jewish wedding; and*
* *Describe the ritual of mourning, the care of a dead body, and the types of graves that were used among the Jews of Bible times.*

A. BIRTHS

The Bible teaches that children are a great blessing from God (Psalm 128:3). Unfortunately, this led many Jews women to feel unnecessary shame if they were not able to bear children, as the stories of Hannah (1 Samuel 1:6), Rachel (Genesis 29:31 ff.), and Elizabeth (Luke 1:7 ff.) all show. The birth of a son was considered a special blessing (Jeremiah 20:15). This was especially true in Jewish families because many mothers dreamed that their newborn child might possibly be the long-awaited Messiah.

When the time came for a baby to be born, a midwife usually assisted the mother in giving birth (Genesis 35:17; 38:28). As soon as the child was born, the umbilical cord was cut, and the baby was immediately washed with warm water and rubbed with salt (Ezekiel 16:4). Then, he was laid with his arms at his side and wrapped firmly with strips of cloth to keep him warm and keep his back straight (Luke 2:7). This arrangement was used for some time, allowing the mother to put the child down and do her work without fear that he would roll over and hurt himself.

If the child was a boy, he was circumcised on the eighth day as God's law required (Genesis 17:9-14; Luke 2:21). This was also the occasion at which the child's name was officially announced (Luke 1:57-60). It was a Jewish custom to name the child after his father or another male relative if he were a boy (Luke 1:59-61), or a female relative if she were a girl. Relatives often gave input on the choice of names. At times, a name was chosen to fit a particular situation (Genesis 4:1) or some peculiar feature of the baby himself (Genesis 25:25), but this custom was fading by New Testament times.

For 40 days after the birth of a boy, and 80 days after the birth of a girl, the mother remained at home. After that period, the mother went to the temple to be purified and to present her child to the Lord (Luke 2:22). She brought a lamb, or, if the family was poor, two pigeons or turtledoves as an offering to cleanse herself (Leviticus 12:1-8; Luke 2:22-24). These were the main customs surrounding the birth of every child. The mother continued to breast feed the child for several years. When the child ended this period of breast feeding and was weaned, the family held a celebration to honor this occasion and to thank the Lord that the baby had survived the critical first years of life. Weaning marked the child's first step towards becoming an adult (Genesis 21:8; 1 Corinthians 3:1-2).

B. BECOMING AN ADULT

Children took the next step towards adulthood about the age of twelve. Old Jewish sources tell us that at that time a boy was expected to begin learning his father's trade or profession more seriously. Also, parents began teaching their children the more exact observances of fasting at that age. Boys underwent training which enabled them to take part in the synagogue and temple service. Luke tells us that Jesus was 12 years old when he went with his parents to the temple (Luke 2:41-52). Men were not considered full adults, however, until they were about 30 years old. They were then considered old enough to be leaders or teachers. Note that Jesus waited until he was about 30 to begin his ministry (Luke 3:23).

C. MARRIAGE

One of the most important and festive occasions in most societies is the engagement (or betrothal) and marriage. Girls were thought to be old enough for marriage when they reached their teens, but often marriages

were arranged much earlier. The main responsibility for choosing a marriage partner rested with the parents. Abraham arranged for a wife to be found for Isaac (Genesis 24:1-4) and Isaac told his son Jacob where to find his wife (Genesis 28:1). Sometimes the feelings of the children were also consulted, but not always. Rebekah was asked if she wished to go to marry Isaac (Genesis 24:58), and Jacob himself selected the woman he wanted to marry from among his relatives (Genesis 29:18). Esau made his parents angry by marrying against their wishes
and taking heathens as his wives (Genesis 26:34-35).

The marriage agreement was finalized with a formal betrothal and engagement between the two people. Usually, relatives or friends acted as go-betweens in the marriage arrangements. The husband normally had to pay the parents of the bride a price for taking her. This was partly to acknowledge that the bride's family had lost her as a laborer, and partly to encourage a lasting relationship (the money would need to be repaid if the woman returned to her family). Jacob worked seven years for each of his two wives (Genesis 29:20), and Shechem offered to pay any price asked to receive Dinah as his wife (Genesis 34:11-12). Deuteronomy 22 makes it clear that the betrothal was considered a final and decisive act. When a women was betrothed, she belonged to her new husband just as though they had already been married (vv. 23 ff.), but the couple was not allowed to live together or have a sexual relationship yet. Betrothal involved a solemn oath and covenant (Ezekiel 16:8). At times, a year or more might pass between the betrothal and the actual marriage. If too much time passed after the engagement, however, people might start to question whether the marriage would really take place (1 Corinthians 7:36).

That a betrothal was considered final and the equal of marriage is seen from the story of Mary and Joseph. After Mary's betrothal to Joseph but before their marriage, an angel announced to her that she would become pregnant and have a son who would be the Messiah (Matthew 1:18). Normally when a girl became pregnant outside of marriage, her family or her betrothed husband would take revenge upon the person who got her pregnant, and the girl herself was often put to death. The Bible praises Joseph as a "just man." Instead of wishing to have Mary shamed and punished for becoming pregnant before she was married, he instead decided to "divorce her quietly" (Matthew 1:19).

The actual marriage festivities had two main parts. First there was a procession during which the groom went to collect his bride, and then there was the wedding ceremony and feast itself. The bridegroom dressed for the wedding better than he had or would dress for any other occasion, with colorful robes and garments and with his head decorated "like a priest" (Isaiah 61:10). After preparing himself, he and his friends set out for the house of the bride. She also was dressed as beautifully as possible. Brides commonly wore wedding ornaments (Jeremiah 2:32), costly garments, jewelry, bracelets, a necklace, a nose ring, earrings, and a golden crown (Ezekiel 16:10-12). In a similar way, the church will be "prepared as a bride beautifully dressed for her husband" when she is united with Christ at the wedding feast of the Lamb on Judgment Day (Revelation 21:2).

The bride's friends then went out to meet the bridegroom and his friends as they approached the house (Matthew 25:1-5). When the bridegroom was in sight, the bride's friends cried out, "Here is the bridegroom; come out to meet him!" (Matthew 25:6). When the procession arrived, the bride and her family and friends were all awaiting the groom. Her father or relatives then gave their blessing to the couple (Genesis 24:59-60). Amid great shouting, dancing, singing, and general rejoicing, the bride and groom then proceeded with their friends to the feast, usually at the bridegroom's house (Jeremiah 7:34).

Then began the second part, the wedding feast or supper (Revelation 19:9). This feast often lasted for days (Judges 14:2). The bride and groom sat beneath a canopy or pavilion during the feast (Psalm 19:5). A special friend or relative was appointed as the "master of the banquet" (John 2:8). He was responsible for greeting the guests properly, overseeing the entertainment, and supervising the supply of food and drink. A great many guests were usually invited (Genesis 29:22; John 2:1-2; Luke 14:16 ff.) and each wore a special wedding garment for the occasion (Matthew 22:11). The guests were seated according to the rank or honor which the bride and groom accorded them (Luke 14:7-9). Wine and food were served in abundance for this once in a lifetime occasion (John 2:3-5; Mark 2:19). The most important element of the feast was its joyfulness – especially as it was shared with friends and family (Matthew 9:15).

D. BURIALS

When a death occurred, wailing began immediately. This wailing spread from house to house until the whole village knew of the death. The women of the neighborhood often continued wailing and lamentation until the body was buried (Mark 5:38). The Old Testament records King David's words of lamentation when his son died, "O my son Absalom! My son, my son Absalom!" (2 Samuel 18:33). At times, professional mourners were also called in (Jeremiah 9:17-18; Amos 5:16). The mourners wore sackcloth or tore their clothes (2 Samuel 3:31) or beat upon their breasts (Luke 23:48) to show their anguish and sorrow. Many of the mourners also cried or wept (John 11:33).

The female relatives quickly washed the body of the dead person, anointed it with spices, and wrapped it in strips of linen. They also placed a small cloth over the face (John 11:44; 19:40). Then the body was put on a mat and carried in procession to the place of burial (Luke 7:14). Because of the heat, the burial took place on the day of death or the following day. The mourners continued fasting until after the burial, and then they had a small feast (Jeremiah 16:7). The closest friends and relatives sometimes continued their mourning for weeks or months.

Because of the rocky soil, the Jews often buried their dead in a natural or man-made cave. This was especially true of the wealthy. Often, the wealthy would own a family tomb for the extended family (Genesis 49:31; 50:13; Judges 8:32). Inside was a larger main room with slabs around the side walls. The body of the newly deceased person was placed on one of the slabs. A year later, a second burial would take place. Because of the normally hot weather in Israel, the flesh would usually be gone by this time and only the bones remained. The tomb would then be re-opened and the bones of the deceased were placed in a cavern at the rear of the tomb with the bones of the other ancestors. In this way, a person was literally "gathered to his people" (Genesis 25:8). It may be that the man whom Jesus said should "let the dead bury the dead" was on his way to such a reburial when Jesus called him to follow him (Matthew 8:21-22). Graveyards were always located outside the city walls (2 Kings 23:6).

LESSON 35:
THE CHURCH SPREADS IN ISRAEL

Before his return to heaven Jesus told his disciples, "You will receive power when the Holy Spirit comes on you; and you will be my witnesses in Jerusalem, and in all Judea and Samaria, and to the ends of the earth" (Acts 1:8).

The rest of the events recorded in the book of Acts and the New Testament letters are a witness that Jesus's disciples carried out his command. Today's lesson outlines how the apostles spread the gospel in Judea, Samaria, and Galilee.

At the end of this lesson you should be able to:

* *Recount how and where the apostles began their preaching and who their leaders were;*
* *Name the apostles who first started preaching in Samaria;*
* *Name the second important Christian center in Israel (besides Jerusalem) and who its founders were;*
* *Identify the first opponents of the Christian church;*
* *Describe how Jewish Christians lived in Jerusalem; and*
* *Explain why and when Jerusalem ceased to be the center of the early Christian church.*

A. THE CHURCH IN JERUSALEM

Even though most of Jesus's first disciples were from Galilee, Jerusalem became the first major center of the Christian church. Shortly before his ascension, Jesus told his disciples to remain in Jerusalem (Luke 24:49; Acts 1:4) and to begin their work there (Acts 1:8). Already before Pentecost, the congregation at Jerusalem had at least 120 members. On the day of Pentecost, 3000 people became followers of the risen Christ (Acts 2:41). Many of those baptized on Pentecost were Jews who did not live in Jerusalem. They had come to Jerusalem from other parts of Israel and from across the Mediterranean to take part in the festival. When they returned to their homelands, they spread the message of salvation far and wide (Acts 2:8-11). The new believers who did live in Jerusalem remained and helped establish the church in Israel. This story about the church's expansion is told in the New Testament book of the Acts of the Apostles, and all the Bible references in this lesson will be to that book unless otherwise stated.

On Pentecost, the Holy Spirit granted a special gift of tongues to the apostles which enabled them to share the message of Christ with people from all over the world. Especially when they preached in new areas, the Lord also gave these early disciples the power to perform miracles to confirm the message that they proclaimed (3:1 ff.; 5:12 ff., etc.). Through these means, the preaching of the Word gained acceptance among more and more people in Jerusalem where the Lord added new believers to his church every day (2:47). In Acts 4:4 we are told that "many who heard the message believed, so the number of men who believed grew to about five thousand." Shortly afterwards, we are told that more and more men and women believed (5:14); the author of Acts also tells us "that the number of disciples in Jerusalem increased rapidly, and a large number of priests also became obedient to the faith" (6:7).

During this time of great growth, Satan was trying to undermine Christ's work in Jerusalem. The temple officials and other Jewish religious leaders arrested and threatened Peter and John (4:1 ff.). Shortly thereafter, all the apostles were arrested and spent time in jail (5:17 ff.). Eventually, the opposition resorted to even more violent means to stop the spread of the Christian church. Stephen was stoned to death (7:59 ff.), and a general persecution of all Christians began in central Judea (8:1). Not until after the conversion of Saul (who later became known as the Apostle Paul) did the church again enjoy a time of peace (9:31).

Sometime after that, however, Herod Agrippa I began to persecute the church. He had the apostle James, the brother of John, executed (12:2). He also arrested Peter, but the Lord sent an angel and rescued him from prison (12:3 ff.). Soon after, the Lord brought about the destruction of Herod so that the church could resume its preaching and teaching in peace (12:19-24).

As the apostles and other Christians continued to spread the good news about Jesus throughout the region, Jerusalem remained the center of the early church. Peter, James, and John were the "pillars" of the church in Jerusalem (Galatians 2:9), but after James's death, new leadership emerged. Jewish resistance to the Romans often ended in riots, which brought violence and unrest to the Christians at Jerusalem (21:27 ff.). When an open and organized revolt finally broke out in A.D. 69, the Roman army took decisive action and crushed the Jewish nation. The Romans destroyed both the city of Jerusalem and the temple. Heeding Jesus's warning of these events (Luke 21:20-24), many Christians fled from Jerusalem and settled elsewhere. After the destruction of Jerusalem in A.D. 70, Jerusalem ceased to have more than an honorary importance in the growing Christian church.

B. THE CHURCH IN JUDEA AND SAMARIA

Soon after Jesus's ascension almost everyone in Jerusalem knew about the synagogue or congregation of Christians which followed the teachings of Jesus (4:16). Once a thriving congregation was firmly established in Jerusalem, the disciples eagerly began the second part of their mission—bringing the gospel to the rest of Judea and Samaria. This work had actually begun already on the Day of Pentecost. Although Acts gives us few details, within a short time, others outside of Jerusalem were being attracted to the ministry of the early Christian church (5:12-16).

The early periods of persecution actually sped up the outreach efforts of the church. The martyrdom of Stephen and the harassment of the church that went with it resulted in Christians being scattered throughout Palestine. They preached God's Word wherever they went (8:1-4). As has happened many times since, the witness of suffering Christians helped bring new people to faith.

The book of Acts does not give us a detailed description of the church's spread into Judea and Samaria. It highlights only a few important events and places during this period of expansion (see Map 26). In Acts 8, Luke relates the story of how Philip brought the Word of God into Samaria. He also preached to an African official from Sudan who had come to worship at the temple. Acts 9 describes Peter's preaching tour to the town of Lydda and the Sharon Plain (vv. 32-35), including the port city of Joppa (v. 36 ff.). From there he continued on and preached to the Roman centurion Cornelius in the city of Caesarea. Caesarea was the second most important city in Israel and was where the Roman governor normally resided (chapters 10-11). The evangelist Philip later lived and worked in this city (21:8), and he probably assisted Paul while he was imprisoned in Caesarea for two years after his third missionary journey (23:23 ff.). For several centuries, Caesarea remained a prominent Christian center in Israel. From this central point, the gospel spread north along the coastline to important cities like Tyre and Sidon (21:3-4; 27:3). Within a few years, the

Christian church and its gospel message had spread through much of the Holy Land.

C. JEWISH CHRISTIANITY

At first much of the Jewish community considered the early Christians in Jerusalem and Judea to be just another new sect within Judaism. Such a view changed when the Jewish religious leaders realized that Christian teaching undermined the foundation for a strictly Jewish church based on obedience to the Law. Christians were perceived as forsaking Jewish customs, dishonoring the temple, and introducing new teachings. The Christians emphasized the justifying and atoning death of Jesus rather than the adherence to Old Testament civil and ceremonial laws. The Jewish sects were not willing to accept such ideas.

Yet, as different as the Christian faith was, in many ways the Jerusalem Christians lived just as the Jews had always lived. They went regularly to the temple to worship and pray (2:46). They took part in the usual sacrifices and purifications (21:23-24). They followed the regulations regarding clean and unclean foods and they kept the Sabbath by attending their own synagogue services. They accepted proselytes (converts to the Jewish faith) into their community as well as Greek-speaking Jews.

For many years Jewish Christians debated about which of their Old Testament religious laws and customs they should continue to observe while practicing their new faith. Some early members still insisted that circumcision was necessary before being baptized into the church. Others demanded that all Christians must continue to obey all the Old Testament laws. The Lord sent Peter a vision to teach him that God wanted his message to be preached to everyone, even those who were unclean according to the old ceremonial laws (10:10ff.). Yet, not following these laws would make future outreach to Jews almost impossible. Therefore, the apostles encouraged Gentile Christians to keep these Jews in mind and observe some of the Old Testament laws (15:20-21).

As a result, the earliest Christians in general enjoyed the favor of the people in Jerusalem (2:47). This was partly due to the generous gifts to the poor given by Christians out of love for their Savior (2:45; 4:34). But, at times, especially when stirred up by their religious leaders, the common people became hostile to Christianity and joined in persecuting the church (cf. 6:13; 12:3; 21:27-28). Thus, Christians spent much time and effort in showing their fellow Jews that they were not degrading God's laws, but that God had a plan of salvation that included all people, not just the Jews.

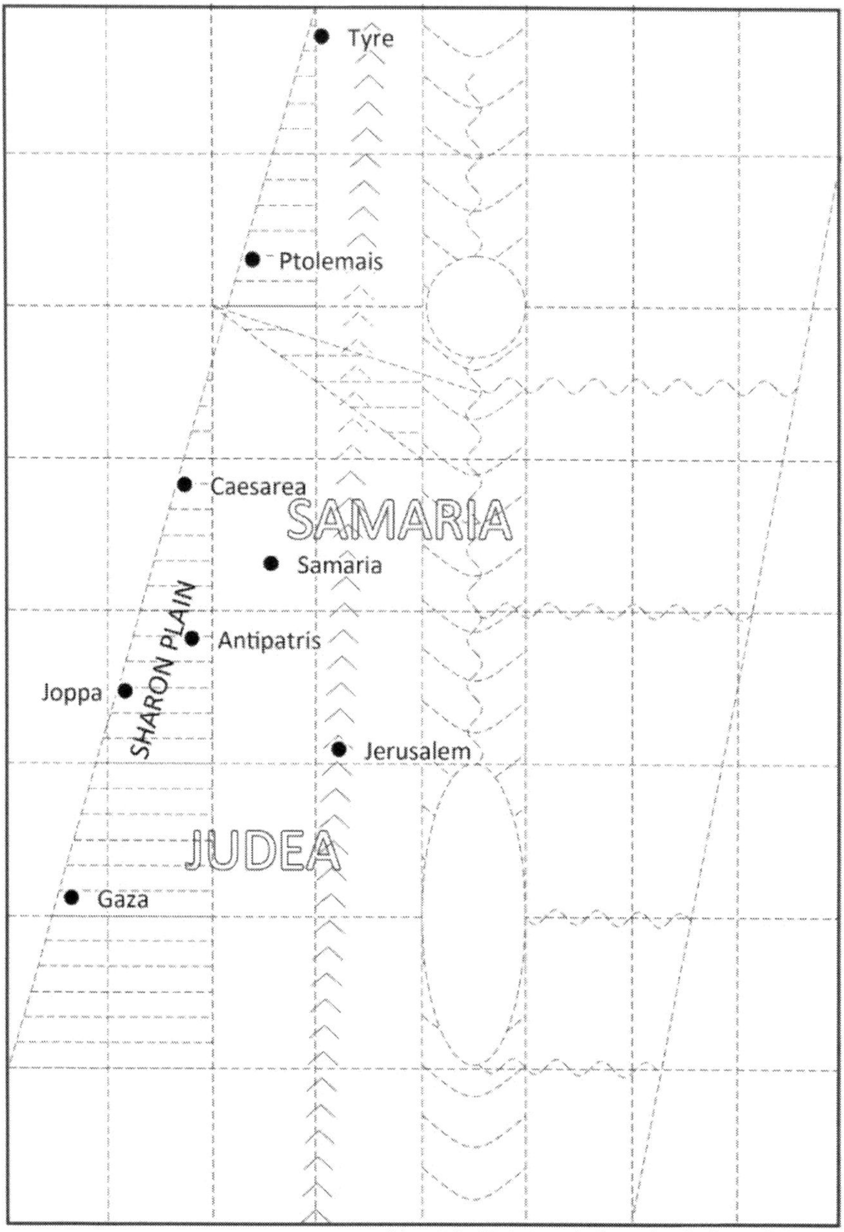

Map 26. Christianity's Spread in Judea and Samaria

LESSON 36:
THE DIASPORA AND THE SYNAGOGUE

In previous lessons, we have studied the geography and history of Israel during New Testament times. We have learned about Jerusalem and its temple and also examined the spread of Christianity throughout Judea and Samaria. Before we describe the spread of the Christian church across the Roman Empire, we first need to take a look at another group of Jews—the Jews of the dispersion or the diaspora. We will study how and why many Jews left the land of Israel to live in other countries and their role in the development of synagogue worship.

At the end of this lesson you should be able to:
* *Describe what is meant by the Jewish diaspora;*
* *Explain what the center of life was for those Jews;*
* *Describe a synagogue, its worship service, and its officials;*
* *Explain what is meant by the term "God-fearing"; and*
* *Describe how the Jewish synagogue was the door through which Paul reached the Gentile world.*

A. THE DIASPORA

The word *diaspora* means scattered or dispersed. This term is often used to describe the Jews scattered throughout the world. The first major dispersion of the Jews was the deportation of the Jews of the Northern Kingdom by Assyria (722 B.C.). A second dispersion was the exile to Babylon of the Jews of the Southern Kingdom (586 B.C.). Many of these Jews never returned to their homeland but remained in foreign lands. Other Jews left the land of Israel voluntarily in search of a more prosperous life elsewhere. The Greek rulers who followed Alexander the Great, especially the Ptolemies and the Seleucids, encouraged Jews to move to other parts of their kingdom. Some of the early Roman emperors also gave special privileges to Jews who lived outside Israel. They were not required to participate in local pagan religious ceremonies. They were allowed to keep the Sabbath according to the Jewish customs and to send money to Jerusalem to pay the temple tax. Both deportations and lenient immigration policies help explain why at the time of Christ large numbers of Jews were living throughout the Mediterranean world.

The extent of the diaspora is evident from the story of Pentecost. Luke tells us there were "God-fearing Jews staying in Jerusalem from every nation under heaven" (Acts 2:5). These Jews had journeyed from foreign lands to beat Jerusalem for the Feast of Pentecost. Luke indicates that most of them understood the language of the country where they lived better than they understood the local Hebrew or Aramaic of Jerusalem. Acts 2: 9-11 gives us a glimpse into the extent of the diaspora:

> Parthians, Medes and Elamites; residents of Mesopotamia, Judea and Cappadocia, Pontus and Asia, Phrygia and Pamphylia, Egypt and the parts of Libya near Cyrene; visitors from Rome ... Cretans and Arabs—we hear them declaring the wonders of God in our own tongues! (See Map 27)

Scholars have estimated that three quarters of all Jews living at that time were living outside of Israel!

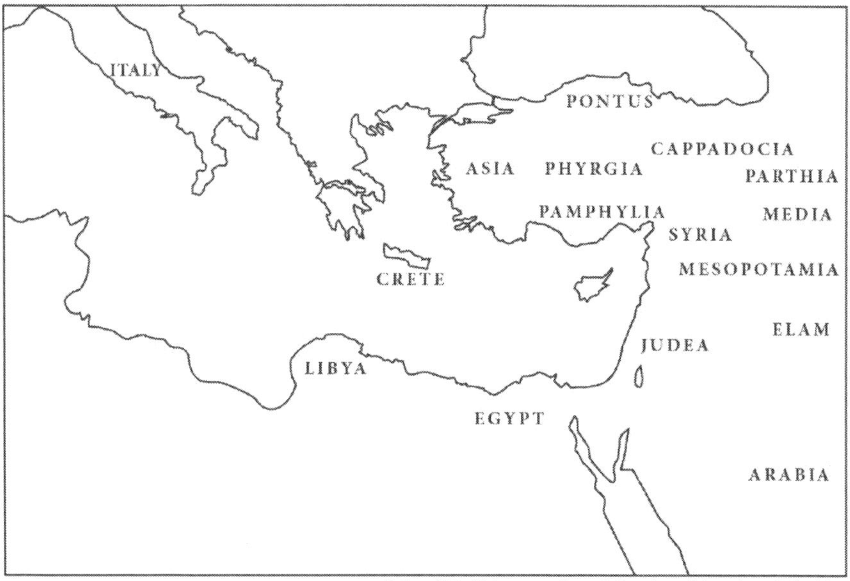

Map 27. Diaspora Jews Mentioned in Acts 2:9-11

B. THE ESTABLISHMENT OF THE SYNAGOGUE

We have already learned that synagogue worship probably dates back to the time of the Babylonian captivity (Lesson 23). From the Gospel accounts, we see that synagogues had become an important part of Jewish life in Israel by Jesus's day. For the Jews of the diaspora, the synagogue was not only the center of Jewish religious life, but of Jewish life in general. Wherever a group of Jews settled, they almost immediately organized a synagogue (Acts 15:21).

It may have been in the diaspora that special buildings were first constructed for synagogue meetings. These buildings were relatively simple, often rectangular in shape. If possible, the building was aligned so the people faced Jerusalem as they worshiped. Large pitchers of water for ceremonial washings stood at the entrance. Often, a picture of a jar of manna was placed above the door. A partition usually separated the men from the women. Most of the people stood or sat on the floor during the service.

At the end opposite the entrance was a raised platform. On the platform was a lectern where the reader stood while reading the Scripture lessons (Luke 4:20 ff.). Behind the lectern were special seats for the most important members of each synagogue. Behind those seats, built into the front wall, was a wooden cupboard in which the scrolls of the Law and the Prophets were kept. Wealthy Jews from the community often paid for decorations on the walls or floor, or a special collection was gathered for such purposes.

Each synagogue had one or more synagogue rulers who were in charge of organizing activities, including the Sabbath services (Luke 13:14; Acts 13:15). An attendant might assist the synagogue ruler by taking care of all the practical matters, such as preparing the scrolls for reading. The ruler also ensured the children were receiving proper instruction and that punishment was properly administered to those being disciplined by the synagogue (2 Corinthians 11:24).

C. THE SYNAGOGUE WORSHIP SERVICE

Traditionally, a minyan (ten adult Jewish men) needed to be present for a service to be held in the synagogue. Over time, a standard order of service came to be used for the Sabbath service. The order of worship consisted of two parts. First came a liturgical section, and then came a time for instruction. The synagogue ruler decided which individuals in the congregation (or which special visitors) conducted each part of the service (Acts 13:15). While somewhat similar to our church services today, the synagogue did not have a single pastor or rabbi lead the service each week. The following description is based on evidence from Jewish sources shortly after the time of Christ.

The liturgical portion of the service began with the Shema – the reciting of three important passages from the books of Moses (Deuteronomy 6:4-9; 11:13-21; Numbers 15:37-41). These passages reminded the people to never forget God's Word. The Shema was followed by a prayer of 18 blessings which asked God to help his people Israel in every possible way. After each prayer, the congregation responded with an "Amen." (In the following centuries, after the majority of Jews had rejected the Christian message, one of the 18 prayers used in Jewish synagogues was changed to include a curse upon Christians.) Psalms were also sung, and additional prayers were offered up to God during the liturgical part of the service.

The instructional portion of the service began with a reading from the Law of Moses, the first five books of the Old Testament. Any literate male adult of the congregation was eligible to do the reading. Readings were arranged so that the whole Law was read through completely in one year. Since the Law was read in its original Hebrew, an interpreter translated it into Aramaic or Greek, verse by verse, so that everyone would understand what was being read. After the reading of the Law, a selection from the Prophets was also read.

Finally, one of the learned men or a distinguished visitor gave a sermon or "word of exhortation" (Acts 13:15). This message was usually based on the two previous readings from Scripture. When the message was finished, the Old Testament blessing used by Aaron (Numbers 6:24-26) concluded the service.

In addition to the regular services on the Sabbath, the Jews also held other synagogue services, especially on festival days. In some areas, services were also regularly conducted on Mondays and Thursdays. During the week, regular classes were held to instruct the children of the community. Many Jews learned to read and write with the Old Testament as their textbook. They were also instructed in the basic teachings of the Old Testament and in how to live properly according to the decrees of the Law. Adult men sometimes gathered at the synagogue to discuss or debate various interpretations of God's laws. Throughout the diaspora the synagogue served as the focal point for Jewish life and helped the Jews maintain their religious and ethnic identity.

D. HELLENIZATION AND THE DIASPORA

Attempts to force Greek customs upon the Jews living in Israel met with only limited success. Although many Jews adopted the Greek language for public conversations and business, the Jews in Israel never became thoroughly Greek or Hellenized. Hellenization was, however, much more common among the Jews of the diaspora. This was only to be expected since they were living and working among Hellenized people.

Many Jews of the diaspora adopted some Greek clothing, customs, and culture. Some became Roman citizens or citizens of the city in which they lived. Jews frequently had both Greek or Roman names as well as their Jewish ones, which allowed them to blend into local society more comfortably. Thus, Joshua was also known as Jason (Romans 16:21), Silas as Silvanus (Acts 15:22), and Saul as Paul (Acts 13:9).

At the same time, however, many Jews retained customs that were strictly Jewish. They continued to circumcise their children, follow the laws regarding eating regulations, and observe the Sabbath. They educated their children in the teachings of the Law of Moses. Some parents even sent their young men back to Jerusalem for training (like Paul of Tarsus). In this way, the Hellenized Jews of the diaspora became a bridge between Israel and the Gentile nations around them. God's Word had always touched the hearts of Gentiles who had been exposed to his teachings, but even most of those Gentiles did not become full converts to the Jewish religion. As a result, they only were given limited access to worship. One reason male Gentiles did not convert as often as female Gentiles may have been that Gentile men were required to undergo the painful act of circumcision. Only a very few non-Jews in the Old Testament period, however, completely adopted the Jewish faith and became full proselytes or converts to Judaism.

The Jewish religion attracted much more interest from outsiders in diaspora communities. Gentiles could easily come to their synagogue services and hear about the God of Israel since the Jewish services were in Greek rather than Hebrew. The Scriptures were also often read in Greek from the Septuagint translation, first made by the Hellenized Jews of Egypt. Also, if some Gentiles wished to become associated with the Jewish religion but did not want to be circumcised or obey all the ceremonial laws, they were accepted as "God-fearers" and were then allowed some status within the Jewish worshiping community (Acts 10:22; 13:50). God-fearers observed the Sabbath and followed the moral precepts of Judaism, but they did not always follow all the Jewish ceremonial laws.

During his missionary travels, St. Paul found a setting for preaching the gospel among such Greek-speaking and Hellenized Jews and those converts to Judaism who worshiped with them. When he arrived in a new city, Paul always started his evangelism efforts in the local synagogue (Acts 17:2). Tragically, the Jews to whom he preached usually rejected the gospel message. Nevertheless, the Gentile converts and God-fearers were far more receptive to the message about Jesus Christ. These converts to the Christian faith were among the founding members of the congregations Paul planted in many cities (Acts 13:42-52; 17:4-12). Thus, the dispersion of the Jews and the ensuing creation of synagogues across the Mediterranean was used by God in his plan for the spread of the gospel.

LESSON 37:
THE GOSPEL'S SPREAD
TO THE ENDS OF THE EARTH

Once the church had been firmly established in Jerusalem and Judea, it was ready to carry out its great commission—to bring the gospel to the very ends of the earth. The dramatic growth of the early Christian church during the time of the apostles will be the focus of this lesson. We will especially highlight the journeys of St. Paul.

At the end of this lesson you should be able to:
* *Indicate how missionary work was made easier by the Roman Empire;*
* *Explain how new ideas spread throughout the Roman Empire;*
* *Name two important cities where the gospel took root and the most important areas where Paul preached on each of his three missionary journeys; and*
* *Tell where the church had spread by the end of this early period of expansion.*

A. THE ROMAN EMPIRE AND ITS RELATIONSHIP TO THE GOSPEL

God chose just the right time and place for the establishment of his church. The Holy Spirit's special gifts on Pentecost helped to jump-start the spread of the gospel. Additional factors, long planned by God, also assisted early Christians in sharing God's Word with the world around them. One key factor was the development of the Roman Empire. During the early Christian period, the Roman Empire extended from the British Isles in the west to the Tigris and Euphrates Rivers in the Near East, and from Germany in the north to the Sahara Desert in the south. In order to maintain control over such a large area, the Romans linked the various parts of their empire with an extensive system of roads. They also patrolled against bandits and rid the Mediterranean of the pirates who had plagued traders for centuries. Thus, travelers and traders were able to journey freely and safely throughout the empire. Because Greek had become the common tongue of most of the major areas of business and commerce in the Eastern Mediterranean (and the Latin language was used similarly in the Western

Mediterranean), people moved easily from one region to the next as they pursued business enterprises.

Rome, of course, was the hub of the empire. It was by far the most important and most visited city in the Mediterranean world at that time. Government officials, soldiers, traders, business men, and sightseers were constantly coming and going to and from Rome. For people, as well as for new ideas, all roads did lead to Rome, and from Rome people and ideas were dispersed and spread to the far corners of the empire. This spreading of people and ideas was to play a vital role in helping the gospel's spread. Thus when the book of Acts ends its story with Paul preaching in Rome (Acts 28:30-32), its readers could assume that from there the gospel would have spread everywhere else where Rome ruled. Map 28 shows the most important land and sea routes from Israel to Rome.

Rome was remarkably tolerant of new religious views. On reason for this tolerance was that Rome had a large and flexible pantheon of gods. Another reason for accepting new religious ideas was the Roman government's policy of allowing subject countries to continue practicing their own customs. Thus, scholars and religious teachers could travel from place to place spreading their views and attempting to gain new converts. This freedom was a tremendous advantage for the early Christian church, for it enabled the apostles and others to freely proclaim the message of Christ wherever they went. Christianity was at first protected by the laws that allowed Jews to worship their God freely. But, the majority of Jewish religious leaders rejected the Christians' claim to be the true Israel (Romans 9:6-8). So by about A.D. 60, government officials began to view Christianity as separate and distinct from Jewish beliefs. Christians were then liable to be arrested and executed as threats to public order. However, instead of destroying the church, persecution strengthened many Christians in their commitment to their Savior. When persecution did come, it merely dispersed Christians further across the empire giving them new opportunities to spread the gospel to even more people.

B. THE FIRST 15 YEARS AFTER PENTECOST

Since the New Testament does not give us precise dates, we can only give an approximate time frame for the 15 years between Jesus's death and Paul's missionary journeys (A.D. 30-45). The first important event during this period was the great Day of Pentecost which took place about A.D. 30.

Following the conversion of 3000 believers on that day, hundreds of these individuals returned to their homelands. They took the gospel message with them and formed Christian congregations wherever they went

The book of Acts provides only a few scattered stories about the spread of the Christian faith during this period. We know, for example, that the gospel was carried into Africa by the Ethiopian eunuch, spread into Samaria by Philip and Peter (Acts 8-10), taken to Phoenicia, Cyprus and Antioch after the death of Stephen (Acts 11:19), and preached in Damascus during the persecutions of Saul. After Saul's conversion, he spent years in his hometown of Tarsus where he may have founded a congregation before going to Antioch to assist the church there (Acts 11:25-26). In a relatively short period of time, the church had spread throughout Judea and Samaria and into regions beyond. However, for most of this period, the gospel was still being preached "only among Jews" (Acts 11:19).

C. THE JOURNEYS OF PAUL

The journeys of the Apostle Paul, from about A.D. 45-60, ushered in a new period of wider mission work. Paul was Gods' chosen instrument to witness to the Gentiles (Acts 9:15). Often called the greatest missionary who ever lived, Paul carried the message of Christ from one end of the Roman Empire to the other. Since Paul was a Roman citizen, he was especially suited for this work (Acts 16:38). Since he was also a Jew, Paul was able to make use of the synagogues scattered across the Roman world as a starting point for his work. As he traveled from city to city and then from country to country, Paul benefited from Rome's highway system and its relatively safe travel by sea. Thus Paul (along with the assistants he took along for training) was able to move quickly from one region to another. In this way Paul spread the Christian faith and planted numerous churches in central and western Asia Minor (modern-day Turkey). Paul went even further by entering Europe and preaching the gospel in Greece. Although there was already a congregation at Rome before Paul visited there, he did assist it with his letter to the Romans.

Though we don't know the full story of Paul's earliest and last mission work, the book of Acts describes the central years of Paul's ministry as consisting in three missionary journeys. Below are listed the key events of each (see also Map 28). All references are to the book of Acts.

Paul travelled with Barnabas on his first missionary journey (Acts 13-14). They started after receiving the Lord's command and blessing and a commissioning by the church at Antioch (13:1-3). They first preached as they traversed the island of Cyprus. Later, Barnabas made a second journey to continue preaching the gospel there (Acts 15:39). Paul and Barnabas then sailed to the mainland of Asia Minor and preached at a number of cities in Southern Galatia – Pisidian Antioch, Iconium, Lystra, and Derbe. These churches were probably the recipient of Paul's letter to the Galatians.

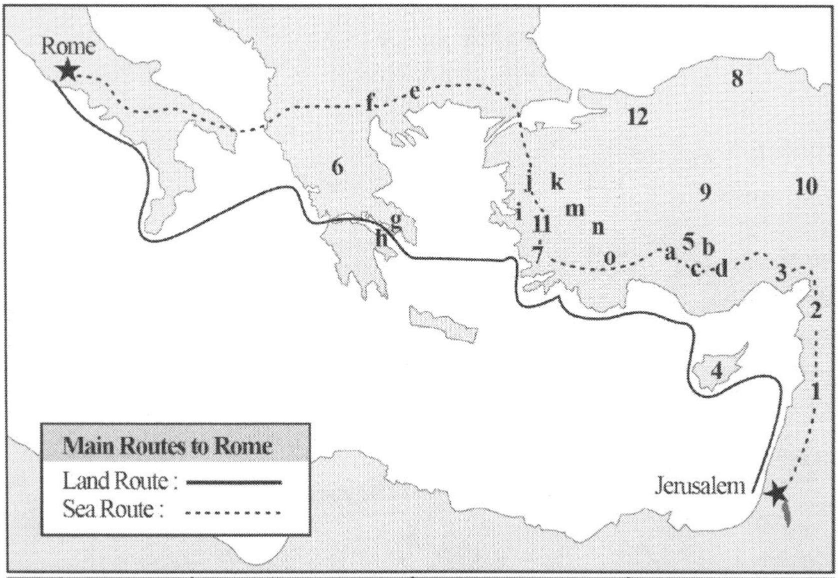

EARLY PERIOD	*PAUL'S JOURNEYS*	*LETTERS OF REVELATION*	*FIRST PETER*
1. Damascus	4. Cyprus	7. Ephesus	8. Pontus
2. Antioch	5. South Galatia	i. Smyrna	9. Galatia
3. Tarsus	a. Antioch	j. Pergamum	10. Cappodocia
	b. Iconium	k. Thyatira	11. Asia
	c. Lystra	m. Sardis	12. Bithynia
	d. Derbe	n. Philadelphia	
	6. Greece	o. Laodicea	
	e. Philippi		
	f. Thessalonica		
	g. Athens		
	h. Corinth		
	7. Ephesus		

Map 28. Early Christian Congregations on the Roads to Rome

After revisiting Antioch and Jerusalem, Paul set out on a second missionary journey (Acts 16-18). This time he was accompanied by Silas. After revisiting the congregations in southern Galatia where he added Timothy to his traveling companions, Paul headed west towards Ephesus. The Holy Spirit intervened, however, and through a vision (16:6-10) directed Paul to the province of Macedonia in northern Greece. There, Paul founded congregations at Philippi and Thessalonica. He then headed south into Greece proper and established a congregation at Berea, preached at Athens, and finally reached Corinth. This important seaport was the capital of the province of Achaea, and Paul spent a year and a half preaching there before returning to Jerusalem (Acts 18:11).

On his third missionary journey (Acts 19-21) Paul again visited congregations that he had previously established, but he also spent two years establishing a congregation in Ephesus, one of the most important cities in the province of Asia. The new church there soon spread the gospel throughout the entire province of Asia (Acts 19:10). By this time Paul had a larger group of helpers and trainees traveling with him (Acts 20:4).

When Paul returned again to Jerusalem, some Jews caused a riot and accused Paul of treason against Rome. Paul was arrested in order to preserve the peace. After two years of imprisonment in Caesarea, he was sent to Rome to stand trial. While he was under house arrest and awaiting trial in Rome, Paul was able to preach to visitors for several years (Acts 28:30-31). From Paul's letters, most scholars conclude that Paul was then released and spent several more years traveling and preaching. Perhaps he even fulfilled his wish of preaching in Spain (Romans 15:24-28). At some point Paul was again arrested and then executed in about A.D. 64. This period of Paul's life is not recorded for us in the book of Acts, which ends with Paul's first arrival in Rome.

D. THE LATER APOSTOLIC PERIOD

The New Testament period lasted almost another 40 years after the conclusion of Paul's journeys. Many stories were told in the early church about events surrounding the other apostles, but the New Testament tells us little. John most likely was the last of the apostles to die (about A.D. 100). After the destruction of Jerusalem and the temple in A.D. 70, the center of Christianity shifted to Asia Minor. By the second century this region had the largest number of Christians.

The apostle John spent the last years of his ministry in and around the city of Ephesus. At the time he wrote the book of Revelation, John had been exiled to the nearby island of Patmos (Revelation 1:9). The book of Revelation contains letters to seven congregations in the province of Asia—Ephesus, Smyrna, Pergamum, Thyatira, Sardis, Philadelphia, and Laodicea (Revelation 2-3). The latter six cities are all within about 150 miles (250 kilometers) to the north and east of the city of Ephesus. Several of these cities became important bases for the continued spread of the church.

Peter's first letter is also addressed to congregations in Asia Minor (1:1), again showing the importance of this region in the growth of the church. Asia Minor remained a Christian stronghold for centuries although the cities of Constantinople, Alexandria, and Rome became more influential than Ephesus as time went on.

LESSON 38: HOW THE BIBLE CAME TO US

The first books of the Old Testament were written almost 1500 years before the time of Christ. The entire New Testament was written in a period of about 50 years which ended just before A.D. 100. So all the books of the Bible were written at least 1900 years ago. In this lesson, we will learn how the books of the Bible were collected and preserved over time. We will also discuss some of the most important translations of the Scriptures.

At the end of this lesson you should be able to:

* *List the three groupings of the Old Testament and the New Testament books;*
* *Name the most important Greek translation of the Old Testament;*
* *Name the most important Latin translation of the Bible and its author and tell how long it was used;*
* *Explain why Martin Luther's translation was important and serves as a good example for other translations; and*
* *Explain why new translations are still needed.*

A. THE WRITING OF THE OLD TESTAMENT

In earliest history, God at times spoke directly to his people. He expected that what he said would be shared among his people and taught to their children (Deuteronomy 6:4-9; Psalm 78:1-8). But that oral transmission of God's Word was gradually corrupted, and many did not hear it or heed it. The truth about himself, his Son, his Spirit, and the redemption that he had brought about for all people all but disappeared by the time of Noah. So, beginning after the flood, God inspired prophets and apostles to write his message down. In this way, his message could be preserved properly over a long period of time, and anyone who wished could hear, read, and study it. Since then, God's people have not only passed on God's message from generation to generation by mouth, but also in written form.

Moses began this process by writing the first five books of the Old Testament under God's direction (Exodus 34:27). When he had finished writing what God had revealed to him "from beginning to end," the book was placed in the Ark of the Covenant so that it would not be lost but would remain a witness to God and his love (Deuteronomy 31:24-26). Joshua continued to write down God's Word following Moses's death (Joshua 24:26), and others continued on with this work after him. Later, God gave messages to various prophets and told them to write these down as well (Isaiah 8:1; Jeremiah 22:30). By about 400 B.C., all 39 books of the Old Testament had been completed. By Jesus's time some Jews had arranged these books into three parts—the Law of Moses, the Prophets, and the Psalms (Luke 24:44). Other Jewish groups divided the 39 books into just two groups—Moses and the Prophets (Luke 16:29,31). These are the expressions used in the New Testament to refer to the Old Testament.

When the Jews spoke of the Law of Moses, they meant the first five books of the Bible—Genesis through Deuteronomy. When they spoke of the Prophets, they were referring to some of the historical books (Joshua, Judges, 1 and 2 Samuel, 1 and 2 Kings) and the prophetical books from Isaiah through Malachi. Finally, when they spoke of the Psalms or the Writings, they were denoting Psalms, Job, Proverbs, Ruth, Song of Songs, Lamentations, Esther, Daniel, Ezra, Nehemiah, and 1 and 2 Chronicles. This three-part arrangement and order still continues to be used by Jews today.

After Alexander the Great's conquests in 333 B.C., Greek became a common language in the eastern Mediterranean. For this reason, and also because the many Jews in the diaspora spoke Greek, the Old Testament was translated into Greek in the centuries before Jesus's life. The most popular Greek version of the Old Testament came to be called the Septuagint. This word means 70, and it comes from a legend which claims that 70 Jewish scholars made this translation. In the New Testament, which was also written in Greek, we find that many of the writers quote from the Septuagint Greek translation rather than from the Hebrew when referring to Old Testament passages. The Greek Septuagint was the version of the Old Testament which most Christians used throughout the early centuries of the church.

B. THE FORMATION OF THE NEW TESTAMENT

Like the books of the Old Testament, the books and letters of the New Testament were also originally written on scrolls—rolls of either parchment or papyrus (Luke 4:17). Inspired by the Holy Spirit, some of Jesus's early followers left written records of Jesus's works and words. They did this to assure other Christians of what had truly happened (Luke 1:3-4). These accounts were also useful in sharing the gospel (the good news about Jesus) with those who had not yet heard about the Savior (John 20:31). For this reason, they became known as Gospels. The letters (or epistles) of the New Testament were written to encourage and strengthen congregations or individual leaders (such as Timothy and Titus) in their faith. These letters were later shared and read in other congregations (Colossians 4:16). Soon, Christians throughout the early church were familiar with these letters of Paul (2 Peter 3:15-16).

During the second century, the 27 books of the New Testament were collected into several groups or volumes. The accounts of Jesus's life were gathered into one group. This volume containing the four Gospels was simply called "the Gospel" or the "four-fold Gospel." The letters of Paul (from Romans through Philemon) became a second grouping. Finally, the other letters (or general letters) and the book of Revelation became a third group, though they were collected a bit more slowly. The Acts of the Apostles was sometimes added to the Gospels (since it continues Luke's story of Jesus) and at other times was added to Paul's letters (since much of it describes Paul's ministry). By A.D. 200, most of the 27 books had been gathered into one of these groupings, and Christian congregations everywhere were using them.

The invention of the codex greatly helped this process. The codex—a book with pages, in contrast to the traditional long scroll—was popularized by the early Christians. Instead of using separate scrolls for each book, all the books in each group could be gathered into a single volume. These smaller codices made the Bible much easier to carry and made it easier to find specific passages quickly.

C. THE LATIN TRANSLATION OF THE BIBLE

After the destruction of Jerusalem, Christianity became more and more separated from the Jews. Gentile Christians made up most of the church by A.D. 100. Knowledge of Hebrew became very rare within the next centuries. Christians used not only the Greek New Testament, but also the Greek translation of the Old Testament. In the eastern Mediterranean, Greek remained the normal language of worship.

In the western Mediterranean, Latin, the language of the Romans, was the common language, so Latin gradually replaced Greek in the churches located from Italy westward. Many people translated various parts of the Bible into Latin which resulted in some confusion. About A.D. 385, the Bible scholar Jerome began revising the current Latin translations in order to produce a more standard translation. Since Jerome knew Greek and Latin very well, and also learned Hebrew, he was the right person for the job. His translation became known as the Vulgate (or *common*) version of the Bible.

Jerome's Vulgate was a very good translation. It was soon widely used and Christians in western Europe continued to use it as their Bible for over a thousand years. Although the Septuagint had included the books of the Apocrypha in the Old Testament, Jerome clearly recognized that only 39 Hebrew books were inspired. He suggested that the Apocrypha be put in a separate section to make this clear. Unfortunately, Jerome's advice about the Apocrypha was soon forgotten or ignored, so the Vulgate continued to include the Apocrypha as inspired Scripture. In the New Testament, Jerome's Vulgate included only those 27 books that were accepted as inspired.

D. FROM JEROME TO LUTHER (A.D. 400-1550)

With translations of the entire Bible available in both Greek and Latin, any educated person in the Roman Empire could readily read and study the Bible for themselves. Another door had been opened for even further spread of the gospel. However, additional new translations were soon needed in other languages for the people who did not know Greek or Latin. Thus began the long tradition of translating the Bible into local languages around the world.

In Syria and Mesopotamia, the Bible was translated into Syriac (a language very similar to Hebrew and Aramaic) in the second or third century. About the same time, translations into several dialects of the Egyptian language (later known as Coptic) were made. In Europe, the missionary Ulfilas (or Wulfila) translated the Bible into the language of the Goths, and the brothers Cyril and Methodius invented the Cyrillic alphabet and used it to translate the Bible for the Slavs of Eastern Europe.

Until the Reformation in the 16th century, the Latin Vulgate remained the most popular translation of the Bible in Western Europe. As Greek and Hebrew fell into disuse, church scholars around Europe came to rely on the

Vulgate almost exclusively. Around A.D. 1300, however, a new university movement arose which also revived the study of Greek and Hebrew. At this same time, new Bible translations began to appear in the new European languages that were developing. John Wycliffe, for example, made an English translation of much of the Bible in about 1380. Others soon followed his lead and produced translations in French, German, and Italian.

Although the Bible was translated into German a number of times over these years, the most important translation was that of Martin Luther. Because Luther translated directly from the Greek and Hebrew texts rather than from the Vulgate, his Bible was more accurate and faithful to the original. Luther's translation could also be easily understood by the common people. Because the printing press with moveable type had been invented in Europe shortly before Luther's birth, Bibles could be printed in mass for the first time and sold much more cheaply than previously. Luther completed his translation of the New Testament in 1522 and of the Old Testament in 1541. With some updating, Luther's translation remains the standard German Bible.

E. BIBLE TRANSLATIONS IN MODERN TIMES

Luther's translation of the Bible into the language of the German people was one of the most important elements in the success of the Reformation. Educated people everywhere could read Scripture for themselves and see the true doctrines that Luther proclaimed. Soon, other reformers were also translating the Bible into local languages. By the 17th century, Bible translations were readily available in most European languages, especially in areas where the Reformation had spread. What was once a book that only the clergy could read had become the possession of the common man. For centuries, the Roman Catholic Church insisted that the Vulgate was the only Bible that could be used in its services and that reading and interpreting the Bible was something better left in the hands of the priests. Because of the translations like those of Luther and because of growing rates of literacy among the population of Europe, the Bible and its message were studied more widely than ever before.

During the 17th and 18th centuries, the modern missionary movement began. Inspired by zeal to bring God's Word to others, missionaries began traveling to Africa, Asia, and the Americas with the gospel message. These modern missionaries also recognized how important it was that the people could hear the Bible in their own language. So the missionaries made this a priority in their endeavors. Usually, one of the Gospels was first to be translated, then the entire New Testament, and finally the whole Bible. Having the Bible available in a language that people could understand was a tremendous aid to bringing the message of salvation to non-Christian peoples.

In 1613, the first Japanese New Testament appeared in print. By 1663, the entire Bible had been translated into an American Indian language. By 1728, Lutheran missionaries had translated parts of the Bible into one of the more important languages of India. The first translation of the Bible into an African language appeared in 1816. Robert Morrison began translating parts of the Bible into Chinese in the early nineteenth century. A group of missionaries and Chinese Christians produced the Chinese Union Version which was published in 1919. It is still the most commonly used translation in China. Today, parts of the Bible have been translated into well over a thousand different languages and dialects. In Africa alone, Bible translations are available in more than 500 different languages.

The work of translating the Bible continues. Along with new translations into additional languages, revisions of old translations continue as well. Languages are always changing. New words and expressions are added, and existing words change in meaning. Therefore, the work of Bible translation will only end on Judgment Day. If we seek to communicate the gospel to all nations, translating the Bible and revising translations will always be an important work.

You have now completed this Bible Background course. We pray that our study of biblical geography, history, and daily life will help you see God's wonderful plan of salvation for the world more clearly than ever. God-willing, you will now put the knowledge you have acquired to work as you continue to study the Bible and teach its message to others. As you do so, you will reap the rewards for all your efforts in completing this course. We pray that the mastery of these Bible basics will enhance your study of Scripture and help you grow in grace and in the knowledge of the greatest of all books – the Holy Bible.

INDEX OF SCRIPTURE REFERENCES

Mark

Luke

21:2　　200

Made in the USA
Middletown, DE
03 February 2025

70032046R00135